Khairulla Usmanov

Cotton cleaner for fine weeds in vertical configuration

Khairulla Usmanov

Cotton cleaner for fine weeds in vertical configuration

The issues of effective cotton cleaning by vertical method are investigated

ScienciaScripts

Cover image: www.ingimage.com

This book is a translation from the original published under ISBN 978-620-7-44996-5.

Publisher:
Sciencia Scripts
is a trademark of
Dodo Books Indian Ocean Ltd. and OmniScriptum S.R.L publishing group

120 High Road, East Finchley, London, N2 9ED, United Kingdom
Str. Armeneasca 28/1, office 1, Chisinau MD-2012, Republic of Moldova, Europe
Printed at: see last page
ISBN: 978-620-7-75707-7

Contents

Kh.S.Usmanov Cotton cleaner from small weed impurities of vertical arrangement. 2023.
The monograph analyses the existing foreign and domestic technologies and equipment for cleaning cotton from fine dust. The issues of creating an effective vertical cotton cleaner on the basis of the development of research methods and modelling of cotton movement in the cleaning section of the cotton cleaner FROM FINE LITTER with vertically and in parallel arranged sequentially moving stake and plate drums are considered.
It is revealed that by increasing the angle of the working area girth with the mesh surface, the number of stakes involved in the process of cotton cleaning increases, by increasing the cleaning effect the natural qualitative indicators of the processed cotton raw material are preserved. The energy and resource-saving vertical cotton cleaner from fine trash, which has passed laboratory and production tests, has been created.

INTRODUCTION

According to the United States Committee on Agriculture (USDA) [1], cotton consumption in Central Asia is increasing significantly as countries implement policies to curb exports of the raw material and support value addition to cotton products.

In the Decree of the President of the Republic of Uzbekistan № UP-60 "On the strategy of development of the new Uzbekistan for 2022-2026" dated 28 January 2022, Annex III "Accelerated development of the national economy and ensuring high growth rates" in the goal № 22 **"Continued implementation of industrial policy aimed at ensuring the stability of the national economy, increasing the share of industry in the gross domestic product and growth of industrial production in 1.4 times" set the task of** "...increasing the volume of production of textile industry**".** [2]**.**

Uzbekistan has switched to full processing of cotton fibre from 2020 by organising cotton-textile clusters and predominantly harvesting cotton using cotton harvesting machinery.

Based on this situation in cotton-textile clusters in cotton processing plants, the following tasks should be solved:

- taking into account the fact that foreign cotton cleaning equipment does not fully meet the needs and requirements of local cotton and textile clusters, modernise the existing equipment and cleaning technologies.
- in connection with the predominant transition to machine harvesting of cotton, it is necessary to create a resource and energy saving cotton cleaner from small weeds.

I-THE CHAPTER

ANALYTICAL REVIEW OF THE STATE OF DEVELOPMENT OF EQUIPMENT AND TECHNOLOGY FOR COTTON CLEANING FROM FINE WEED IMPURITIES

1.1 Study of the technique and technology of cotton cleaning from small weed impurities abroad

In the USA (Fig. 1.1), gravity inclined drum cleaners (9) with hot air heating from the drying system are used in the technological scheme to remove fine sap. The cleaner can be used as a pneumatic separator. After cleaning from small weeds, the cotton goes to the cleaner (10), for cleaning from twigs, sashes and other large weeds. The cotton undergoes sequential cleaning on two cleaning sections consisting of serrated drums and grates; the third regeneration section is designed to return flyings to the general cotton flow.

The following drying and cleaning system is equipped with an Impact shaker cleaner (12) mounted under the inclined gravity cleaners (9). This cleaner is only available from Continental Eagle. It is designed for cleaning cotton with high blockages. Cleaning takes place as a result of the interaction of the cone drums with a series of toothed discs forming a rotating grate. Like all inclined coarse litter cleaners, the Impact cleaner is equipped with a regeneration section. The whole complex of the above equipment is located in one production building with a minimal transport area. During the drying and cleaning process, the cotton is in continuous contact with hot air, which ensures moisture extraction at all transitions. The temperature of the heat transfer medium and cotton is controlled by a sensor and controllers, which ensures that the raw material is fed to the gins with a stable moisture content of 6 per cent. The listed disassembly, drying and cleaning technology is equally suitable for both saw mill and roller mill. All drying and cleaning equipment is available in two modernisations, with different widths of working bodies. If the plant capacity is higher than 23-30 bales per hour (fibre), an additional processing, drying and cleaning line is provided in the listed layout with the installation of a second separator on the distribution conveyor, which is equipped with a hinged valve above each gin.

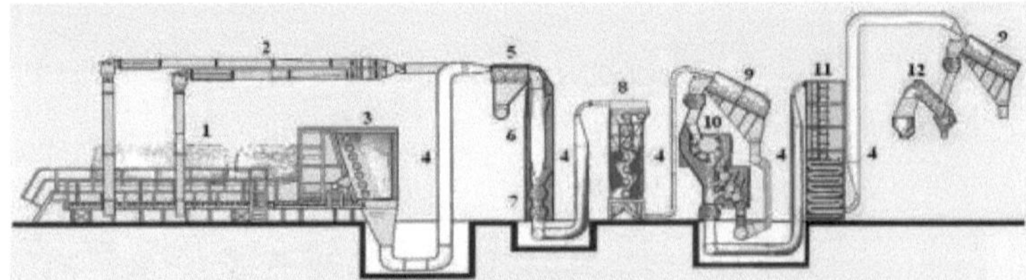

Fig.1.1 Technological sequence of equipment for drying and cleaning of

cotton

1.Cotton module; 2.Feeder auto regulator; 3 Disassembly section;
4.Pipelines; 5.Vig "J" separator; 6.Hopper; 7.Vacuum valves; 8.
Vertical flow dryer;
9.Fine screen cleaner; 10.
Coarse screen cleaner; 11.Tower dryer; 12.Coarse screen cleaner.

The six-drum cleaner of fine sap (Cleaner - 96" and 120") is all-metal, produced in two variants, width -96" (2438 mm.) and width -120" (3045 mm.), have six cone drums (Fig.3), installed in one row with a slope of 30:35 o to the horizontal. Under the drums there are also ponderous grates with a clearance of 5-7 mm. It is produced in different variants: there are models with regeneration saw drum, there are models used as a separator-cleaner and other types. The main purpose of these developments is to increase the productivity of the processed cotton and the cleaning effect of the machine with maximum preservation of the technological parameters of cotton.

The cleaner (Fig.1.2) works in the following sequence: cotton through the pipeline through the nozzle (1) in a mixture with air enters the first cone drum (2). The drums rotate in the direction of flow, so the cotton moves loosening on the surface of the cone drums until it reaches the last sixth drum and passes to the bottom part (under) of the drums. The cotton now moves in the opposite direction. As the rotating drums with their stakes loosen the cotton on the surface of the pond grates (3). Due to the impact and centrifugal force of the rotating drums, the fine grit falls out through the lumen (opening) of the pond grates and cleaning takes place. The cleaned cotton from under the first drum falls into the trough (4) for the next processing. The separated sorghum from the hopper (5) is discharged by the sorghum transport.

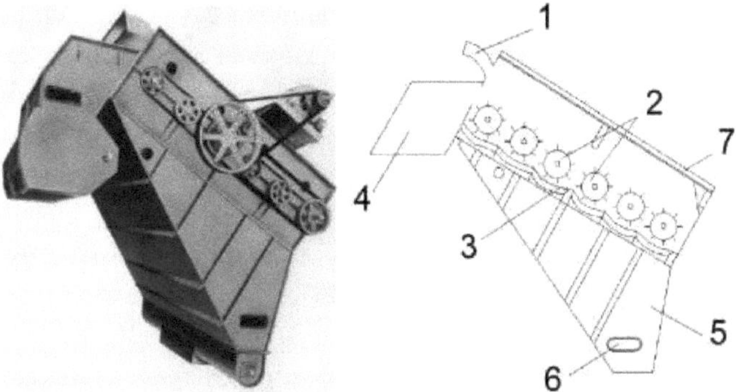

Fig.1.2 Appearance and process diagram of the six-drum cleaner (Cleaner - 96'and120')

1.Inlet pipe; 2.Stake drums; 3.Prudential grates; 4.
Raw cleaned cotton outlet tray
; 5.Sap
hopper; 6.Inspection window.

At cotton ginning plants in India, Bajaj Steel Industries Ltd (Fig.1.3) [3, p.-42-51] are used for cleaning of fine weed impurities.

Inclined type fine weed cleaners are effective cotton cleaners. They usually consist of 4 to 6 staking drums arranged in series, which pull the cotton through a mesh surface to clean it. The beater drums rotate anti-clockwise at the same speed.

Due to the rotation of the stakes drums, cotton is dragged along the entire length of the drum and due to the impact and shaking action, fine weeds are extracted from under each stakes drum. The cleaning surfaces are made in the form of a mesh surface or grates. The weed impurities released through the mesh surface or grate are removed from the machine by the weed auger.

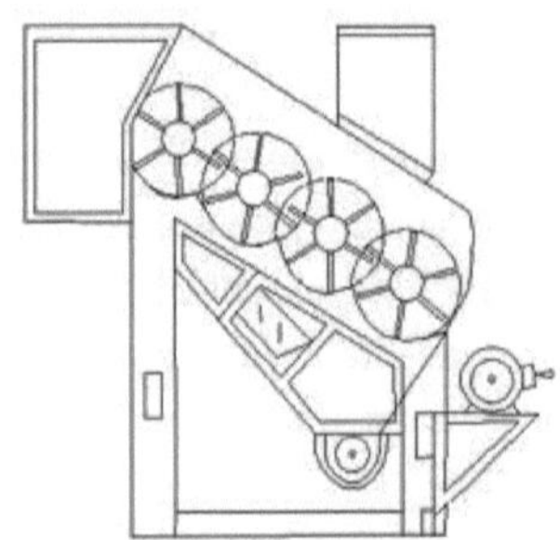

Fig. 1.3 Inclined cleaner manufactured in India (Design by Bajaj Steel Industries Ltd)

The AVI Ginning Machinery or Kawaadi horizontal fine weed cleaners (Fig. 1.4) effectively clean cotton from various kinds of weed impurities such as mineral impurities and fine organic impurities. The principle of operation of the horizontal cleaner is similar to that of the inclined cotton cleaner for fine weed impurities. The number of drums is determined by the cleaning requirements and can vary from 6 to 10 drums and all staking drums are arranged horizontally one after another. The main working bodies of the Kawaadi cleaner are the feed rollers, the main cone drums, the control device and transmission mechanisms, electric motors. The feeder consists of two feed rollers. Under the rotating feed rollers there is a 100 mm diameter main stake drum. Stakes are arranged in four rows and there are 15 stakes in each row.

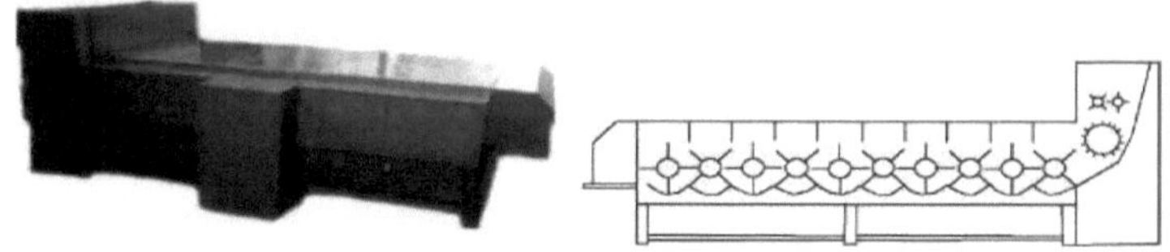

Fig.1.4 Horizontal cleaner manufactured in India (AVI Ginning Machinery design)

The cleaning drums are 150 mm in diameter and have six rows of 140 mm long pegs. There are 15 pegs in each row and the distance between them is 85 mm. The peel drums are placed parallel to each other.

In China, cotton cleaners are installed in the drying and cleaning sections of the cotton mill. This task is also fulfilled by feeder-cleaners installed on each gin. The main manufacturer of cotton equipment is Swan AC with a factory and technical centre in Shandong Province. Swan has developed more than 30 new technologies. Cotton cleaners from small weeds in the People's Republic of China have a design identical to the American machines [4], for example, the MQZK -2400 cleaner (Fig. 1.5) has two cleaning sections, which cleans from small and large weeds. It has two motors, one for the peel drum and the other for the saw drum. Cotton with the help of a guide roller (1) enters the inside of the cleaner on the surface of the peel drums (2). Being loosened by the peel drums it is directed to the outermost drum (3) where it changes its direction moving under the peel drums.

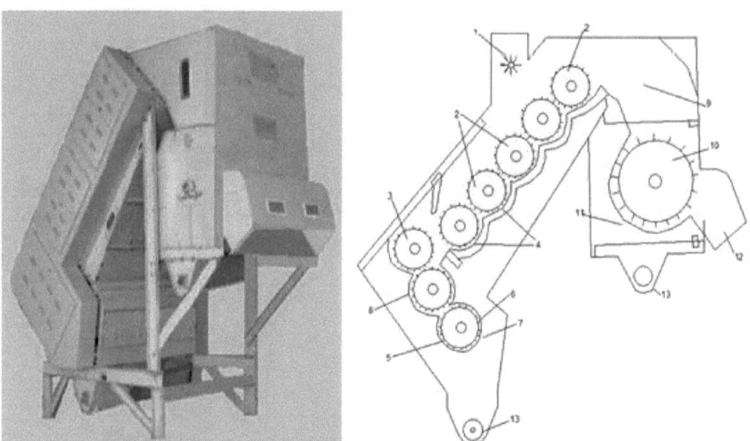

Fig.1.5 External view and cross section of the MQZK-2400 cotton cleaner

1.Guide roller; 2.Stake drum; H.Edge drum; 4.Bar grates; 5.Regeneration serrated drum; b.lapping brush; 7.Spike grate; 8.Brush drum; 9.Shaft; 10. Spike cleaning drum; 11.Grid surface; 12.Tray; 13.Screw conveyor.

The cotton is dragged through the bar grates (4) by centrifugal force and by striking the drum stakes the weed impurities are separated through the gap of the bar grates. The weight of the drums is the same

The drums are of the same design and rotate in the same direction with the same speed, due to which the cotton loosening is transferred from one to the next, in the course of the process, and the cotton is subjected to cleaning simultaneously moving towards the transition shaft (9).

The cotton to be cleaned is now transported to the next cleaning section, where a large diameter cleaning drum (10) is installed. The cotton is once again cleaned from small impurities. Since a mesh surface (11) is installed under the drum, small weeds are extracted through its openings and the cleaned cotton is discharged through a tray (12) to the next process (machine).The individual cotton flies in the mixture of weed impurities released under the influence of staking drums (2) in the first section, are caught (regenerated) by regeneration saw drum (5) and with the help of brush drum (8) are removed from the teeth of saw drums and transferred (ejected) into the main stream of cleaning to the staking drums. The separated weed impurities fall into the weed hopper and are discharged from the machine by a screw conveyor (13).

Figures 1.6 and 1.7 show the designs of the PRC cleaners.

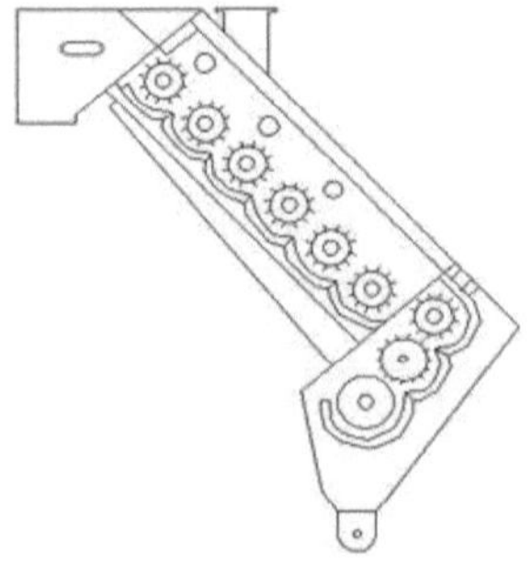

Fig.1.6 PRC MQZH-15 inclined cotton cleaners

Mark	Capacity kg/hour	Power consumption, kW	Overall dimensions, mm	Weight kg
MQZX-15	13000-15000	22	4460x3344x3726	5700
MQZX-12	10000-12000	18,5	4010x3344x3726	5200
MQZX-IO	8000-10000	15	4010x3344x3726	4500
MQZX-17	6000- 7000	11	2810x3344x3726	3500

Fig.1.7 Characteristics of the MQZX series inclined cotton cleaners

The analysis of foreign cleaners of cotton from fine weed impurities shows that the development of research on improvement of the cleaning process was carried out by means of horizontal and inclined combinations of working bodies (stake drums). In addition, in all foreign designs of fine sap cleaners the cleaning sections do not work efficiently enough, as the working angle of contact between the staking drum and the mesh surface does not exceed the value 90°.

1.2 Analysing domestic cotton cleaning techniques and technology to remove weed impurities

The reforms implemented by the Government of the Republic over the years of independence have significantly improved the quality of cotton fibre and increased the profitability of the cotton industry [5], but cotton cleaning technology has not changed significantly over several decades.

The removal of fine weed impurities from cotton is an important process in the primary treatment of cotton, which has a significant impact on subsequent processes such as ginning and fibre ginning. In case of poor quality cleaning of fine impurities, they turn from passive impurities into active impurities, which in turn complicates the fibre ginning process.

In fine sap cleaners for the extraction of fine weed impurities from cotton, cleaners consisting of staking drums (or staking screws) working in combination with mesh surfaces or grates are used.

Fines cleaners are axial and direct-acting, depending on the way the working bodies impact the processed cotton. In cotton cleaners of axial method of action cotton is fed from one narrow section of the machine and moving along the entire length of the working organ (auger), as a result of repeated impact on it, the process of moving along the axis of the working organ, cleaning and unloading in the final section of the machine.

In straight-acting cotton cleaners, the cotton is fed into the working zone along its entire length simultaneously and moves in a plane perpendicular to the longitudinal axis of the drums. In this case, the cotton to be cleaned is subjected to a single impact from the same working elements (pick drum).

An example of axial-acting fine pickling cleaners is the screw-type fine pickling cleaner 6A-12M. (Fig. 1.8), previously widely used for cleaning cotton from small weed impurities in cleaning shops of cotton ginning plants. It is characterised by a high cleaning effect, very simple in its design and reliable in operation. The disadvantages of this cleaning machine include multiple rotational impact on the cotton, which leads to burning of the fibre, and subsequently to the appearance of defects in the fibre.

The cotton entering the cleaning machine is separated into two independent streams and is subjected to the rotating upper staking augers (3).

The stake auger is a 400 mm diameter screw of a conventional conveyor used for cotton transport, with feather-welded stakes that protrude 75 mm above the circumference of the feathers.

The auger spikes, arranged in a helical line, loosen the cotton, toss it up and gradually move it to the opposite end of the machine.

In the process of movement and constant shaking of the cotton, weed impurities are extracted from it and fall through the grates or screens (5) that form the auger chute.

Having been cleaned in the upper parallel augers, the cotton flows through vertical connecting shafts (4) into the same lower augers (6), which again loosen, shake and move the cotton free of debris in the opposite direction - to the discharge opening (7).

The weed impurities extracted through the grates (screens) of the upper and lower sections of the augers fall into the hopper and are removed from it by the weed conveyor (11).

Due to the impact of the stakes on the cotton as it moves, the release of weed impurities is very intensive.

Each lint is in the machine for an average of 30:35 seconds; during this time it is repeatedly exposed to the auger stakes. All weed that has fallen through the perforated nets is transported through the sloping walls of the weed hopper into the weed auger and out of the machine

Technologically, all models of fine litter cleaners work in the same way: when the cleaning drums impact the cotton slices and bales are repeatedly impacted against the mesh surface, the cotton is preliminarily loosened, thus creating conditions for the separation of weed impurities, which are gradually sieved and removed through the openings of the mesh surface.

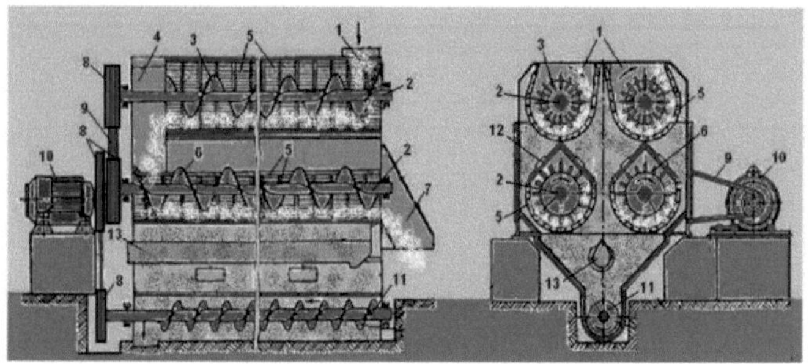

Fig.1.8 Technological scheme of screw cleaner 6A-12M

1.Shaft; 2.Screw shaft; H.Screw; 4.Connecting shaft; 5.Spike (mesh) surface; b.Bottom auger; 7.Unloading holes (chutes); 8.Pulleys; 9.V-belt; 10.Electric motor; 11.Threshing auger; 12.Inclined plane; 13.Pipework for suction of dusty air from the threshing hopper.

To clean cotton of manual and machine picking from small weed impurities with humidity not more than 14% in cleaning shops are installed cleaning machines of direct flow method of action of 1XK or СЧ-02 mark. These cotton cleaning machines are also used as part of flow lines in cleaning and drying and cleaning shops of cotton plants. Cotton cleaning machine 1XK includes a stake section and a cotton cleaner from small debris. Stake section consists of two stake blocks, racks, tray and hopper. The fines cotton cleaner includes a power supply unit, a stake unit, racks, and a hopper for the outlet of the fines. For ease of maintenance, modern 1HK cleaners are equipped with sections of two staking drums of EN.178 brand (Fig. 1.9).

The assembly of four sections results in an eight-drum cleaner 1HK (Fig. 1.10) [12].

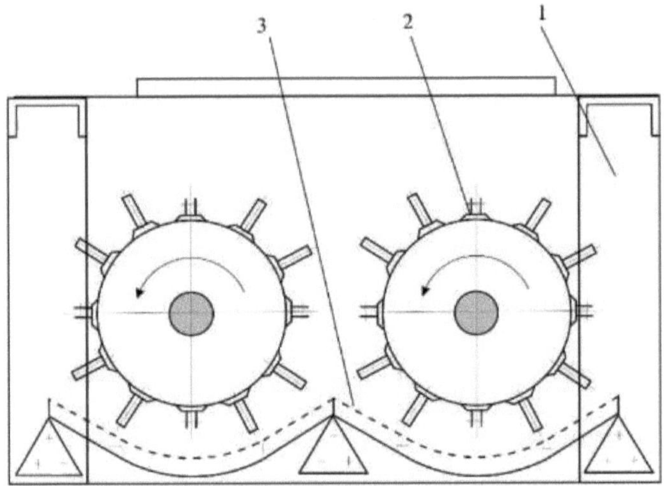

Fig.1.9 Cleaning section of the EN.178 brand
1-frame, 2- stake-and-plank drum, 3- mesh surface.

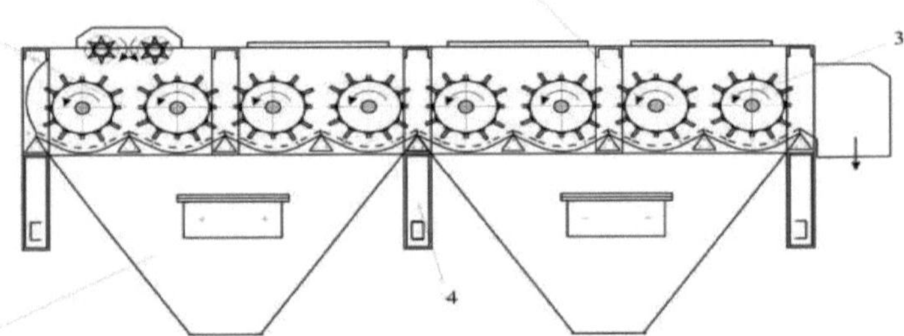

Fig.1.10 General view and schematic diagram of the 1XK fine scum cleaner
1-column-plate section of EN. 178.01 (with feed rollers);
2, 4 racks, 3-flap-plate section of EN. 178.02, 5- hopper.

The practicality and convenience of the sections EN. 178 is that there is a possibility of arrangement and obtaining any number of sequentially installed sections for cleaning from small weed impurities. These sections are also used in cotton ginning units of UCC.

Cleaner of cotton from small debris СЧ-02 (Fig.1.11) is installed in flow lines of cotton processing in cleaning shops of cotton ginning plants. The technological scheme of the cleaner SCh-02 is identical to the cleaners of 1XK brand.

Cotton from the machine interfaced according to the technological process enters the feeder on the feed rollers (1), which feed it evenly to eight successively installed ginning drums (3) mounted on the frame (8) of the housing. Stake drums mix the cotton, trap it on grates (4) and move it to the last drum and further into the tray (6), from which it is fed to the mating device according to the technological process. The fine impurities released during the movement of the cotton on the grates are removed through the hopper (5).

The disadvantage of these cleaning machines is their high metal intensity and the presence of high damage to the cotton during cleaning due to the horizontal arrangement of the stake sections.

In straight-through cleaners, the ripping drum can be a stake, slat, toothed, stake-slat, and sub-drum screens can be woven with cells 10x10 mm stamped with oval holes 6x30 mm, or 6x50 mm, stamped with round holes with a diameter of 10 mm. and grids made of wire with a diameter of 6 mm and a gap between them of 5 *mm* to create the necessary area of the living section of the grid.

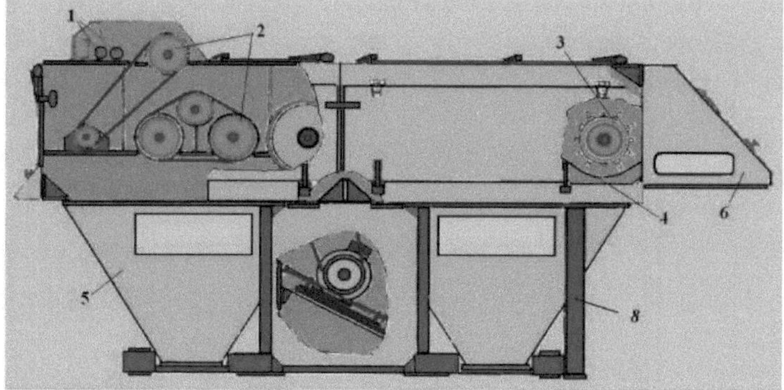

Fig.1.11 General view of the cleaner СЧ-02

1. Feed rollers; 2. Drive; 3. Spike drum; 4. Spike grate (perforated screen); 5. Scum hopper; 6. Tray;

The main working organ of fine sorghum cleaners is a stake and slat drum [6] is a prefabricated structure (Fig. 1.12), which consists of a shaft (1), discs (3), thin sheet lining (2) and slats (6), of which there are eight stake and four paddle slats. The thin sheet liners are bolted together and form 4 slat rows and 8 stake rows around the perimeter of the drum. The slat connections of the liners create an air flow during cotton cleaning, which favours the efficient separation of fine weeds from the cotton being cleaned.

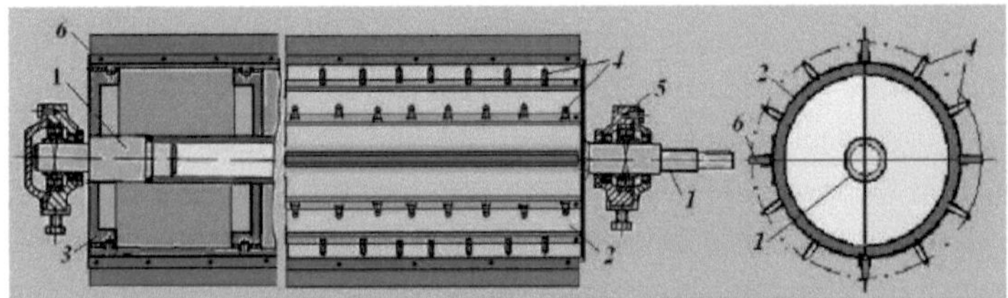

Fig.1.12 Construction of a stake and plank drum

1.Support journal (drum shaft); 2.Cladding (drum shell) 3. Disc flange ; 4.Stake rail; 5.Bearing; 6.Bar;

It has been found that along with the design of the drum, the cleaning effect is influenced by the design of the mesh surface. In the cotton cleaning industry, a perforated sub-drum screen is used (Fig. 1.13). In modern designs of fine sap cleaners, stamped mesh surfaces with the size of holes 5x50 mm. and the location of the large axis of holes perpendicular to the movement of cotton in the cleaner are used.

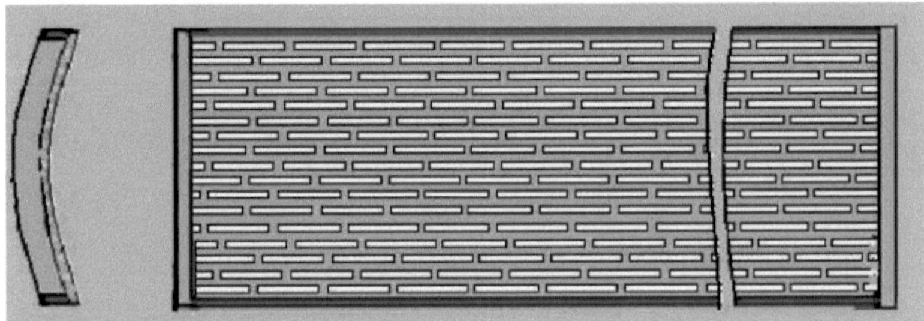

Fig.1.13 Perforated sub drum mesh

The conducted researches on studying the existing technology of cotton cleaning from fine sap and studying the work of cotton cleaning machines with ripping and cleaning drums of the above mentioned constructions (Drum speed V =3,8 m/sec) in combination with sub-drum nets of different types makes it possible to establish the following:

1) optimum circumferential speed of the cleaning and loosening drum 9:10 *m/sec;*

2) the clean surface of the under-drum screen contributes to a consistent cleaning effect;

3) in all operating cotton fines cleaners there is counter rotation of adjacent drums and there is significant damage to cotton (especially in the processing of low grades) due to the abrupt change of direction of cotton movement in the next cleaning section in the course of the process, which has a negative impact

on the quality and significant loss of the final product, as due to the impact impact there are short fibres in the released weed impurities.
4) In the cotton cleaning machine the process of cotton transport is carried out due to unidirectional movement of cone drums, where in the zone between two adjacent drums, which move towards each other cotton particles, which are subjected to significant impact from the cone adjacent drums.
5) During transport, the linear velocity of the staking drums is V1 = 9 m/s, and during oncoming movement the cotton particle is impacted with a velocity of v_2 = 18 m/s (the velocity of the oncoming adjacent drum is added). This results in significant fibre and seed damage.
6) In fine cotton cleaning machines, both horizontal and inclined cleaning sections do not work efficiently enough, as the working angle of contact between the cone drum and the mesh surface in all machine designs does not exceed 120° .

1.3 Review of research on cotton cleaning of fine weed impurities

In cotton fines cleaners, the main cleaning module consists of a vehicle with a different design of headset, which ensures the loosening and further transport of the cotton flow, and a mesh surface, through which the process of fines separation is carried out. In practice, a set of such modules in different variations of installation forms the overall profile of the cleaner. Also in the modules there are other various factors that contribute to the effective separation of fine impurities such as air flows and the presence of slats, which have a certain impact on the process of cotton cleaning from fine impurities [7- p.31-32]

In order to improve the process of cotton cleaning from small weed impurities theoretical studies were conducted by such scientists as Boltabaev S.D., Samanadarov S.A., Levkovich B.A., Jamalova M.M., Huseynov V.N., Tyutin P.N., Lugachev A.N., Sosnovsky Y.S., Sultanov A., Juraev A.J., Hakimov Sh.S., Madumarov I.D., R.I. Ruzmetov on various directions, such as the study of the impact of the impact of the impact of the headset of the drum stakes on the transported cotton, force parameters of interaction, trajectories of movement of cotton particles on the drums, studies of the pile-draining ability of mesh surfaces and the influence of temperature regimes on these processes.

In the field of cotton cleaning from fine weed impurities, experimental studies were mainly based on the optimisation of technological parameters of the cleaning section, feeding zone and introduction of technical innovations to improve the efficiency of the cleaning equipment. According to the results of the experiments, the speed modes of the stakes-plate drums, the profiles of the headsets, the gaps and divisions providing the efficiency of the module and the whole machine were studied and optimised. Researchers Sosnovsky Y.S. and Sultanov A. in their works studied the possibility of using forced air flow in the cleaning zone when cleaning cotton from small weed impurities, but these works were not continued due to the imperfection of the proposed designs.

The studies of M.M. Jamalov, V.N. Huseynov, P.N. Tyutin, A.E. Lugachev, A.H. Bobomatov are devoted to the study of the efficiency of using mesh surfaces, where the influence of changing the orientation of mesh cells on the picking ability was studied and positive results were obtained, but there was a sharp decrease in the reliability of equipment operation due to the slaughter of mesh cells by fibrous material, as well as the influence of mesh surface oscillations on the efficiency of cotton cleaning from small weed impurities.

Below is an analysis of the research carried out by scientists on various directions of improving the technology and design of cleaning machines.

In the work of S.D. Boltabaev [8- p. 156] devoted to the improvement of the process of cleaning of cotton machine harvesting is concluded that frequent re-lapachivaniya and re-shipment of clogged cotton lead to a strong activation of sap, complicate cleaning and reduce the yield and quality of fibre.

Studies conducted by Budin E.F. [9 - p. 146] theoretically and practically established that in order to reduce seed damage during cotton cleaning the circumferential speed of saw drums should be set not higher than 7 m/sec.

G.I.Miroshnichenko [10, - p. 124], studying the issues of increasing the efficiency of fine dust cleaners, suggested to evaluate the work of mesh surfaces by the efficiency coefficient of the "live" section.

Musakhodzhaev Z.M. [11 - p. 11] analysing the work of screw cleaners leads to such disadvantages as the location of stakes on the edge of the screw surface and in the plane perpendicular to the axis of the screw is only one stake. Consequently, the cotton located in this plane is exposed to only one spike per one turn of the screw. At the same time, only a small portion of cotton can be caught by the spike and thrown over the screw, and the rest of the cotton moves along the chute without being subjected to loosening and re-cleaning, due to the insufficient number of spikes. In addition, the screw cleaner has a space between the turns of the screw. A considerable part of the cotton is in this space, the part closer to the screw axis is not loosened and is transported without contact with the screening surface, which reduces the cleaning effect of the machine.

V.A.Bogomolov [12 - p.171] in his researches proves that cotton cleaning on grate-saw and screw cleaners more than two times is not expedient, as some reduction of fibre clogging is overlapped by increase of technological defects.

a) as a result of three-fold cleaning of cotton on CHCH-3M, in comparison with two-fold cleaning, the content of sap in fibre decreases by 0,12% (abs.),

and the number of technological defects increases by 0,15%;
б) multiple cleaning of cotton contributes to the growth of the number of fibre defects, which after threefold cleaning on CHCH-3M almost doubles (7.2% against 3.8% in the initial one).
A similar pattern is observed for multiple cleaning of cotton on screw cleaners.
A.L. Sapon in his research [13 - p. 135] proves that the regulated process of primary cotton processing, which played its positive role at a certain stage, has now exhausted itself. The disadvantages inherent to the regulated process are: dusty air in the premises, low degree of automation, change of flows by cross-section and density, lack of flexibility in adjusting the intensity of cleaning, additional, not related to the process of primary processing of cotton, mechanical impact on cotton, impossibility to implement this process in sets of equipment.
Abduazimov Sh.H. [14 - p.135] in order to reveal the mechanism of cotton cleaning from small weed impurities developed dynamic and mathematical models of impact interaction of cotton particles with working bodies. Mathematical models of movement of a cotton particle with a weed particle, carried by a stake, along the screening surface are obtained.
Lugachev A.E. [15 - p.99] studying the work of fine sorghum cleaner 1HK came to the conclusion that when moving cotton between adjacent stakes drums, under certain conditions there is a precedent of excessive cluttering of operational space in the zone of "reception-transmission" of cotton on the drums to the guides and will limit manoeuvrability in the transport of cotton particles in the working area of the cleaner, which in turn will cause a decrease in technological reliability of the system.
Bobomatov A.Kh. [25 - p.115] on the basis of numerical solution of the problem of oscillations of an elastic plate at constant perturbation from cotton the laws of oscillations of an elastic plate are determined. The equations of oscillatory motion of the elastic plate of the mesh surface at different modes of cotton impact are obtained.
In the studies of I.D.Madumarov [17 - p.181 it was determined that the amount of active weed impurities in the composition of cotton of the selected sample from the bunt is 0.28%, after the first and second drying drums the amount of active sap of cotton increases to 0.5%, and after the first and second line of the cleaning unit UCC decreases to 0.15%.
The model of the process of separation of weed impurities from the composition of two or more elastically interconnected flys is proposed. The author has established that to increase the efficiency of the cleaning process and to preserve the original natural properties of fibre the optimum values of fibre temperature

are 45-50 ^{o}C and fibre humidity 5,5 ÷ 6,0%.

In the above mentioned studies this element of fine sap cleaners has not found sufficient development and further study of this issue, in our opinion, is of particular importance for the creation of resource and energy saving technology of cotton cleaning from fine sap.

D.A. Usmanov [18] developed a prototype (production) of a four-drum cleaner for cleaning cotton from small weed impurities. Its technological scheme is shown in Fig. 1.14.

The cleaner includes a shaft 1, feed rollers 2, stake-plate drums 3 and screening surfaces 4 located under them, having two pockets 5 and spigots 6 each, a weed auger 8, and a chute 7 for discharging cotton from the machines.

The cotton coming out of the feeding shaft is caught by a pair of feeding rollers and is transferred to the staking drum, which throws it into the pockets, and then, hitting the falling cotton is dragged along the mesh surface 4 and contributes to the separation of litter from it. The cotton is thrown on the second staking drum and in the same way passing the other drums leaves the machine by means of the tray 9. The separated weeds in the lower part of the drums fall into the weed augers 8, and in the upper part they are removed from the machine to the aspiration system by the forced suction 10.

Due to the fact that the cleaner of this design uses two AO-2-42-4 electric motors of 4.5 kW power with 970 rpm for the drive, it is energy-intensive in relation to the operating fine dust cleaners of 1HK grade and difficult to maintain.

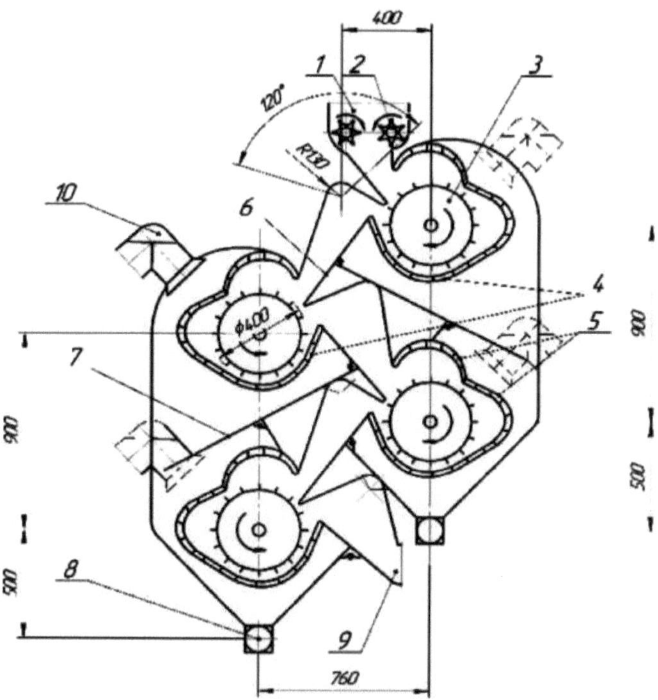

Fig. 1.14 Four-barrel cotton cleaner for the removal of small weeds

II-THE CHAPTER

THEORETICAL STUDY OF COTTON CLEANING TECHNOLOGY FROM SMALL WEED IMPURITIES

2.1 Calculation of force effects on the cotton flywheel at the technology of horizontal method of cleaning from fine litter

As it is known, cotton cleaning is carried out on fine and coarse sap cleaners, which are composed in a certain sequence. The issues of the theory of air flow influence on the efficiency of cotton cleaning from fine sap and movement of cotton fly on the ginning surface, the influence of cotton friction coefficient on the ginning surface, as well as studies of cotton particle scattering fan at the direct-pass method of material transfer are considered in detail in the works of Professor A.E. Lugachev [15].

To the problems of improvement of the process of cleaning from fine sap and increasing the angle of girth of the cone-plate drum with a mesh surface is devoted the work of researcher A.H.Bobamatov, who developed a design of cotton cleaning unit with girth of the drum with a two-pocket mesh surface in 2500 (Fig.2.1).

In this modernised design, the cotton to be cleaned is cleaned by the stakes and slat drum, which captures the cotton with the stakes and drags it along the surface of the mesh surface with curved elastically mounted plates, due to which there is an intensive separation of small weed impurities. The curved elastically mounted plates are made of 65G steel, and in order to increase the cleaning effect they are located along the entire working length of the mesh surface, and the gap is set so that the cotton seed passes freely between the stakes and the mesh surface equipped with curved elastic plates. During the cotton cleaning process, the curved elastic plates allow to give the cotton to be cleaned a dynamic movement, in which small weeds are more intensively released. The vibration allows the cotton to be cleaned more intensively to release fine impurities [19 - p.233].

However, due to the complexity of the design and the presence of frequent downhole situations, this technical solution has not found further application.

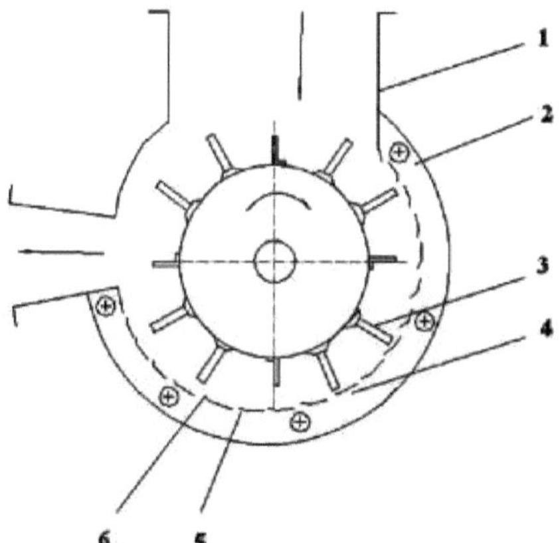

Fig.2.1 Fibre material cleaner

1-housing; 2-pulley drum; 3-pulley; 4-pulley zone;
5-elastic element; 6-holes.

Studies in this direction, carried out in the United States of America [20-27] showed that they were mainly conducted in such directions as the creation of technologies with increasing multiplicity of cleaning, with innovative developments to improve the working bodies of fine sap cleaners, the wide introduction of automation of process control, theoretical and practical study of the influence of warm air on the processes of cotton cleaning.

The analysis of the above studies shows that the process of effective cotton cleaning from fine weed impurities is very complex and there are a number of aspects in this direction that require further theoretical research.

Theoretical description of the mechanism of separation of fine weed impurities from fibrous material is time-consuming, as there is no possibility of precise determination of the strength of adhesion of weed impurities to fibrous material in its various layers. Also the issues of trajectory of cotton and weed impurities during cleaning are not sufficiently fundamentally studied, and applied models describing effective methods of cotton cleaning from small weed impurities have not been created. In our opinion, further theoretical research in these directions is relevant.

On 1XK cleaning machines, the cleaning sections are inefficient because the operating angle of contact between the cone drum and the mesh surface is 100° (Fig.2.1.2 a).

In addition, in the area where the cotton is transported from one drum to the next, there is considerable damage to the cotton (especially in the case of low grades) due to the abrupt change of direction of the cotton in the next cleaning section in the process.

To maximise the utilisation rate of the mesh surface, a vertical working unit of the fine sorghum cleaner has been developed, which operates on the principle of vertical cotton cleaning with sequential movement of the cotton flow without counter impacts (Fig.2.2 b).

28- c.64-65]. In laboratory tests of the cleaner the length of the working part of the cone drum was made in the size of 300 mm. The rotating parts were mounted on cantilever shafts. The mesh surfaces and other elements were fixed on the body of the laboratory unit.

For visual observation of the cotton cleaning process, the front part of the laboratory unit is covered with transparent organic glass, which allows video recording of the cotton cleaning process. When the experimental unit was set up, the arc of girth of the mesh surface of the cone drum reached a maximum value of 210° (Fig.2.3).

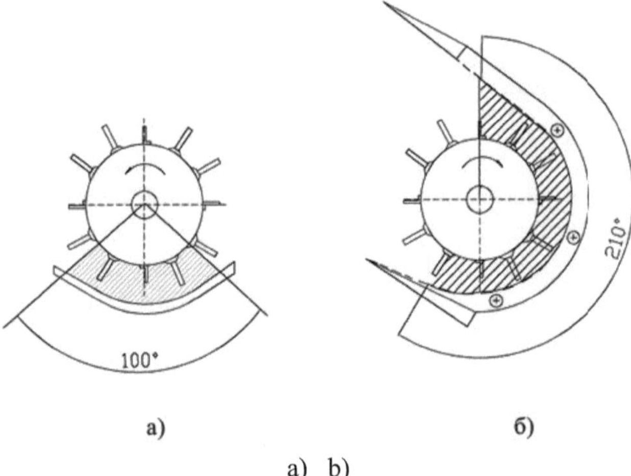

a) b)

Fig.2.2 Schematic diagram of the construction of the cone drum and mesh surface in the existing designs of CCS cleaners a) and the proposed one b)

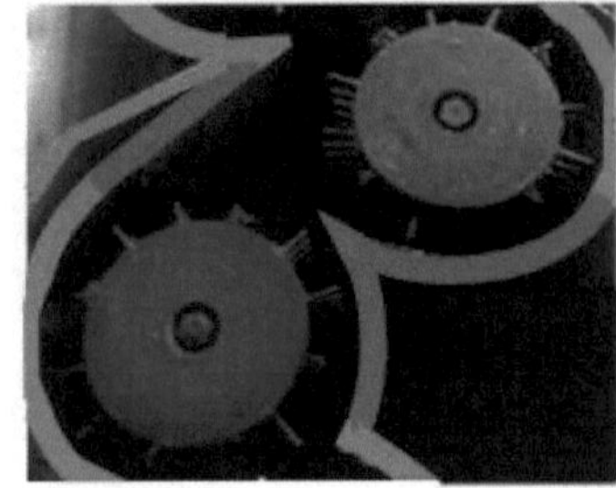

Fig.2.3 Stake drum and mesh surface designs in the existing UCC cleaner designs a) and the proposed one b)

In order to study the influence of forces in the zone "stake drum - bale - mesh surface" on the natural qualitative indicators of processed cotton at horizontal movement of cleaned cotton we have carried out the following calculations. At rotation of the stake drum the cotton boll moves along the mesh surface as a result of which centrifugal force arises [29 - p. 60-63].

In the opposite direction to the rotation of the snare drum is the frictional force of the fly against the mesh surface.

The rotational speed of the cone drum is 450 rpm. Centre distance between two drums is $d = 420$ mm, drum radius $R = 200$ mm, grate height $l = 50$ mm and weight of one flyer is accepted . $m = 6$ г [30 - с.69-74, 31 – с/74-78].

According to Newton's second law, the resulting acceleration of the fly is directly proportional to the applied force and inversely proportional to its mass $F = ma$ [F] =кг·м/с^2.

We assume that the fly on the mesh surface moves uniformly, and then determine the forces acting on it (Fig. 2.4).

We calculate the force acting on the flyer of mass $m = 6$ g from the side of the cone drum of radius R [32 - p. 10742 - 10747].

To carry out calculations, we denote the rotation frequency of the cone drums by v, the drum radius R, the linear velocity of the cone drum" , which is determined by the following formula.

$$\upsilon = 2\pi v R \qquad (2.1.1)$$

It is known that the rotational frequency of the snare drum is equal to:

$$v = 420 об/мин = 420 об/мин = 7{,}0 об/мин \qquad (2.2.2)$$

From here we find the linear speed of rotation of the snare drum

$$\upsilon = 2 \cdot 3{,}14 \cdot 7{,}0 \cdot 0{,}20 = 8{,}792 \text{ м/с} \qquad (2.2.3)$$

A bale of cotton of mass *m is* subjected to a force from the side of the cone drum and a frictional force between the bale and the mesh surface.

Determine the force acting on the cotton flyer from the side of the pick drum using the following formula:

$$F_1 = ma - F_{ish} = ma - \mu mg = m(a - \mu g) = m(\upsilon^2 / R - \mu g) = \\ = 6 \cdot 10^{-2}(81/0{,}20 - 0{,}25 \cdot 9{,}8) = 2{,}4H \qquad (2.2.4)$$

here:

and - coefficient of friction of cotton against steel surface ;
m - mass of cotton fly ash,
g is the acceleration of free fall
a - acceleration of cotton flying.

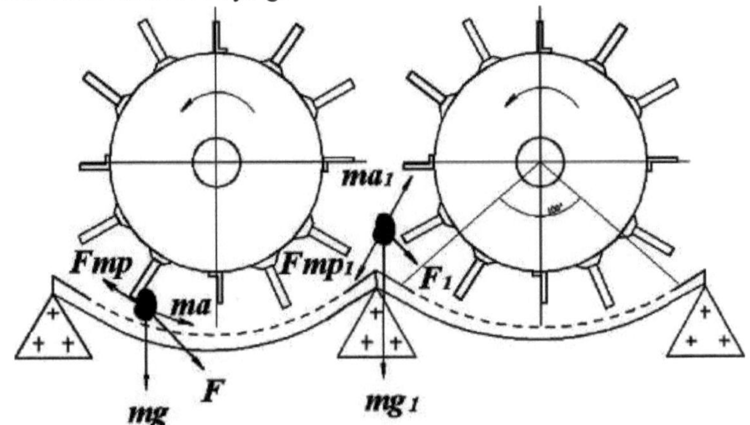

Fig. 2.4 Schematic diagram of forces acting on a cotton flyer during cleaning from fine weed impurities.

After the first peel drum, the cotton boll flies out to the second peel drum, which has unidirectional movement. At the counter movement of the second drum, the cotton boll striking receives the following force effects: *mg* - force of gravity, F_1 *F* - force acting from the side of the second cone drum, equal in magnitude to the force of the first drum [40 - p.233].

After the first snare drum, the cotton boll flies out to the second snare drum, which has unidirectional movement. At the counter movement of the second drum, the cotton bolls impacting receives the following forces: *mg* = *mg- force* of gravity, *F* = *F1-* force acting from the side of the second cone drum, equal in magnitude to the force of action of the first drum.

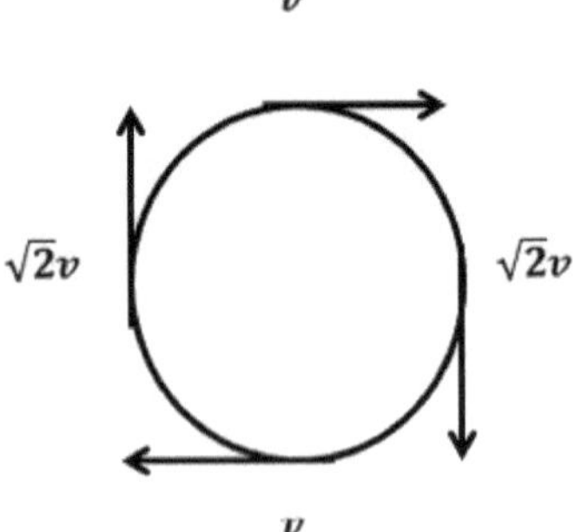

It is known that in a rotating drum the velocity distribution is different (diagram). According to this distribution, after fly fly from the first drum at the point of contact with the second drum, the rotation speed of the cone drum is equal to $\upsilon = \sqrt{2v}$.

1) on the[1/4] part of the rotating drum: $\upsilon = \sqrt{2v}$
2) on the[1/2] part of the rotating drum: $\upsilon = v$
3) on the[3/4] part of the rotating drum: $\upsilon = \sqrt{2v}$
4) at full rotation of the rotating drum: $\upsilon = v$

In the second drum, the acting forces on the fly will be equal.

$$F_1 = ma_1 + mg_1 \qquad (2.2.5)$$

Since the value of *mg1* is small, the solution to this equation is:

$$F_1 = ma_1 = m \cdot \frac{\vartheta_1}{R} = \frac{2m\vartheta_1^2}{R} = \frac{6 \cdot 10^{-3} \cdot 2 \cdot 81}{0{,}2} = 4{,}8 \quad \text{H} \qquad (2.2.6)$$

The results obtained show that a single fly of cotton during its cleaning from small weed impurities during its movement from one picking drum to another is subjected to a total force of 7.2 N. The repeated impact of this force leads to high mechanical damage to the seeds and the formation of short fibres, which are removed from the machine together with weeds [33 - p. 34-3]. Multiple impact of this force leads to high mechanical damage of seeds and formation of short fibres, which are removed from the machine together with weed impurities into waste [33 - p. 34-37].

Studies by American scientists Patil.P.G., Anap G.R., Arude V.G. and Carlos B. Armijo, Kevin D. Baker, Sidney E. Hughs, Edward M. Barnes, Va Marvis N.
in this direction confirm our results, they prove that the number of mechanical influences on cotton significantly damage its natural qualitative indicators. Also, in order to achieve these goals, they conclude to minimise in cotton cleaners from small weed impurities the number of stake drums by improving the assemblies and layout of machines [34, 35- p.158-165].

The calculations show that the existing technology of cotton cleaning from small weed impurities is inefficient and to a certain extent negatively affects the natural qualitative indicators of processed cotton.

According to the results obtained on the fly cotton when cleaning it on the cleaners of fine sorghum when moving from one to the second drum acts a force of 7.2 N.

In turn, this factor is the cause of fibrous waste, which allows us to conclude that it is necessary to modernise the horizontal arrangement of cleaning units with a transition to vertical technology of cotton cleaning.

2.2 Calculation of force effects on the cotton flyer in the vertical fines cleaning technology

At cotton ginning mills and cotton-textile clusters nowadays for cotton cleaning from small weed impurities serial eight-drum cleaners of СЧ-02, 1ХК marks and stake blocks of ЕН.178 mark in a set of flow line UHK, developed in the late 80s of the last century, are used.

The author proposes to modernise this equipment, in particular, in co-authorship he has developed a variant of vertical cleaning of cotton from fine sap, which allows by successive movement of staking drums to increase the arc of the mesh surface, thereby significantly increase the cleaning effect of the machine [36 - p.31-35].

In order to study the character of cotton fly ash movement in the section of cotton cleaning from fine weed impurities, we carry out a number of theoretical calculations. We assume that due to the small distances between the pick drum and the opposite wall of the mesh surface, the velocity is a constant value and varies insignificantly.

After impact, the fly loses its speed and falls on the second drum under the action of its gravity, thus changing the position of the fly in the composition of the flow, because when flying out of the first cone drum on the second drum, it changes the cleaned side, on the opposite side (Fig.2.5). When hitting the mesh surface of the second drum, which in the primary version was made at an angle of 30° to the horizontal axis of the first cone drum (Fig.2.6 a), part of the cotton flow returns back. This occurs due to the fact that after hitting the mesh surface the cotton flow falls to the left of the centre of the staking drum and the incoming staking row, capturing the incoming cotton, does not have time to completely drag the entire mass of cotton into the cleaning zone. To eliminate this negative phenomenon, we have chosen the angle of inclination of the mesh surface relative to the horizontal axis of the first staking drum equal to 45° (Fig. 2.6 b).

These mesh angle parameters virtually eliminated the amount of cotton that was recycled back during processing.

Further research on improving the shape and parameters of the mesh surface will allow to eliminate this negative phenomenon completely. In the scheme of the vertical cleaner with four staking drums their frequency of rotation from top to bottom starting in the first from 390 rpm to 420 rpm. This allowed to eliminate bogging situations, provided uniform uninterrupted transport of cotton on the staking drums and significantly reduced the amount of cotton return to the previous staking drums (Fig.2.7, 2.8).

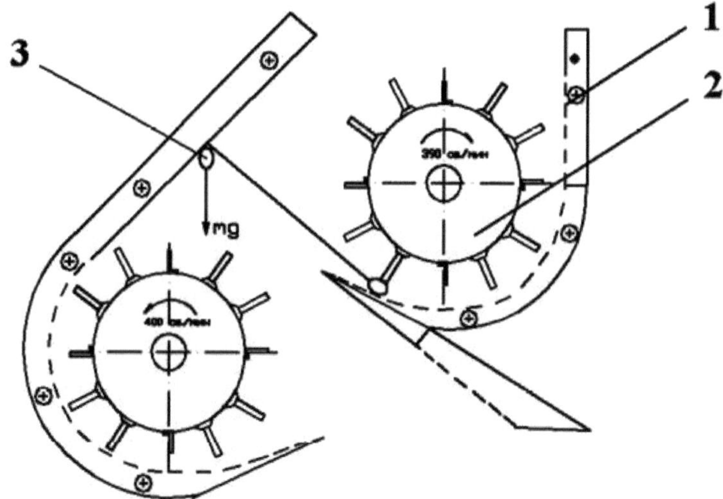

Fig 2.5 Trajectory of the cotton fly between the cleaner's cone drums when cleaning from fine litter

1. Mesh surface; 2. Stake drum; 3. Single flyleaf in the cotton flow composition

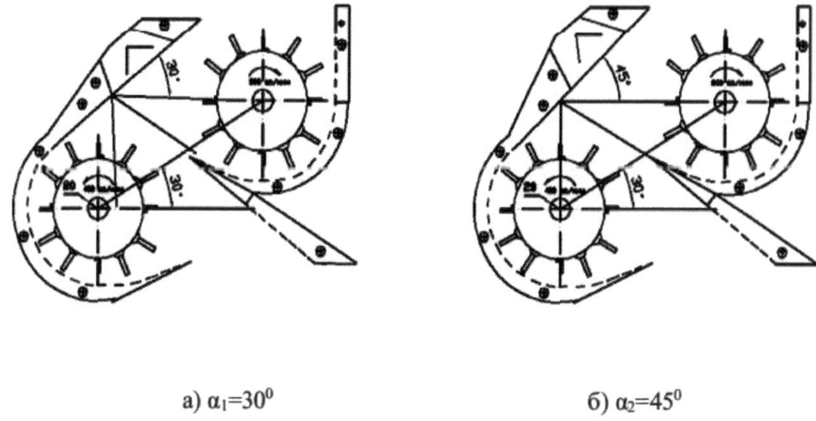

a) $\alpha_1=30^0$ б) $\alpha_2=45^0$

Fig.2.6 Schematic of changing the angle of inclination to the horizontal axis of the first stake drum and the area of cotton falling on the second stake drum

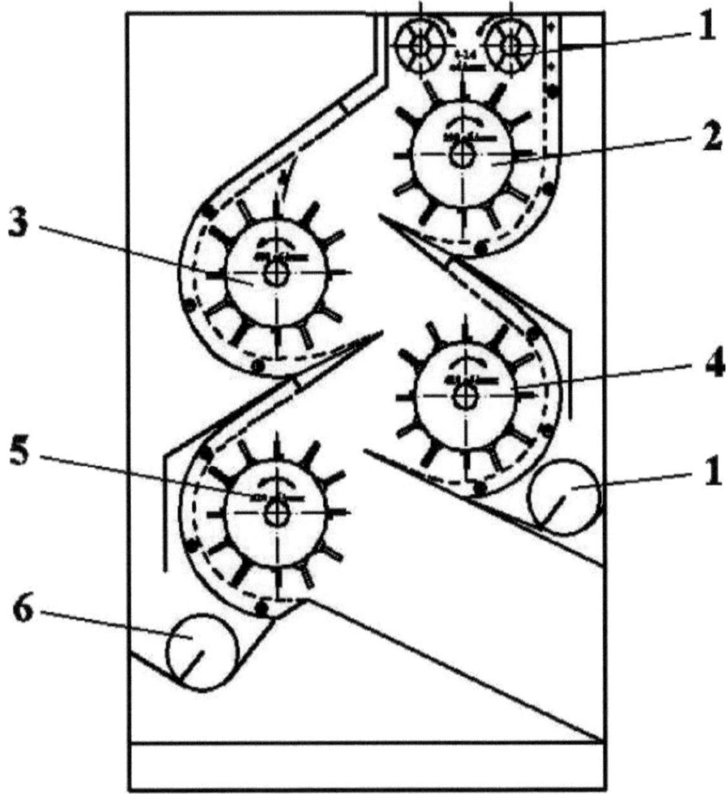

Fig. 2.7 Schematic diagram of a four-drum vertical fine cotton cleaner

1.Feed rollers; 2. First stake drum; 3. Second stake drum;
4. Third stake drum; 5. Fourth stake drum;
6. Weed augers.

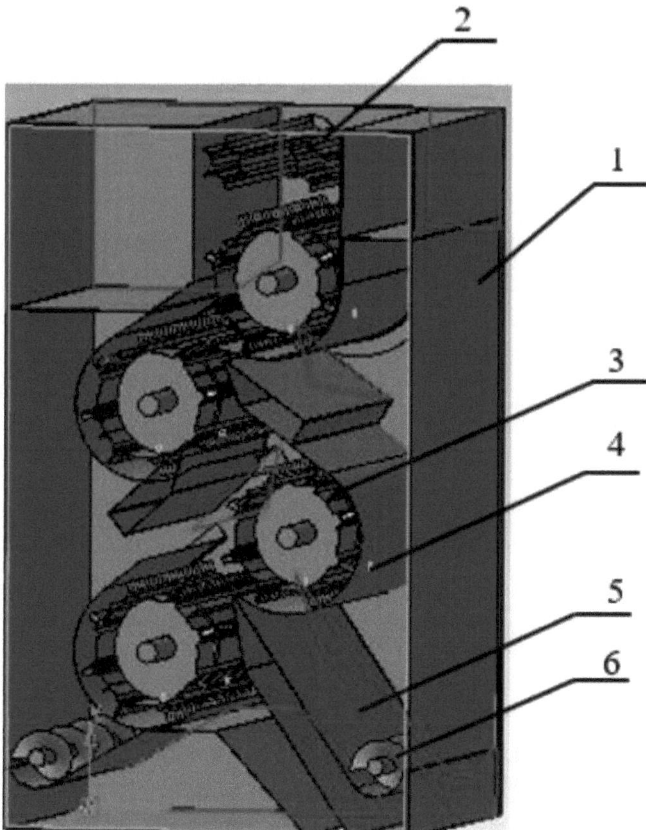

Fig. 2.8 Three-dimensional diagram of a four-drum vertical fine sod cleaner
1 - casing; 2. - feeding rollers 3 - feed drums; 4- mesh surface; 5 - guide; 6. Grinding augers.

Another constructive feature of the vertical cotton cleaning machine is the angle of girth of the mesh surface of the cone drum. If this index at horizontal arrangement does not exceed the value a_{∂} = 100° , in the modernised vertical cleaner this index has increased up to a_{∂} = 210° , that is higher on 2,1 times the useful area of net surfaces of serial cleaners of fine weed 1HK. If at horizontal technology of cotton cleaning from fine weed impurities the coefficient of efficiency of "live" section does not exceed the value ^=0,25^0,30, in the proposed variant ^= 0,58^0,60.

The next stage of the calculations was to calculate the force of impact of the cotton fly against the wall of the mesh surface after flying out of the stake drums of a vertical cleaner consisting of four stake drums.

After the cotton bale enters the zone of cleaning from mecklich weeds, it is subjected to the following forces: the centrifugal force *F-z*, the friction force *Etr* and the gravity force *P from the* side of the stakes of the staking drum (Fig. 2.9). The rotational speed of the peeler drums is different for each of the four peeler drums starting from the top of the peeler 390,400,410,420 rpm respectively.

v1 - 390 *rev/min; v2* - 400 rev/min; v_3 - 410 rev/min; *v4* - 420 rev/min. (2.2.1)

Distance between the axes of drums is $d =^{400}$ mm, the radius of the drum with a mesh surface $R =^{200}$ mm, the length of stakes is $l =^{50}$ mm, the average mass of one fly of cotton we take equal to $m =^{6}$ g.

To determine the forces acting on the cotton fly, we calculate the linear velocity of each staking drum using the formula °= *2^R^V* (2.2.2)

R-radius of the stoke drum, v- rotation frequency of the stoke drum, n=3.14. Further we determine the value of forces acting on the cotton fly when cleaning it from small weed impurities at vertical cleaning technology on each stoke drum.

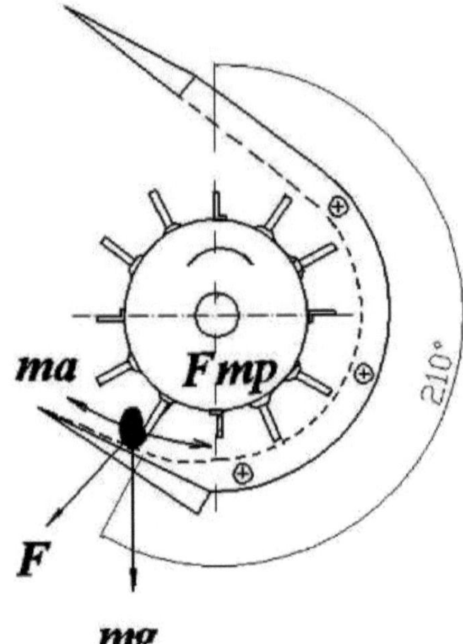

Fig.2.9 Schematic of the forces acting on the cotton flyer at vertical cotton cleaning scheme

We carry out calculations for the first bell drum:

$$\nu_1 = 390об/мин = 6{,}5об/сек \quad (2.2.3)$$

$$\upsilon_1 = 2\pi R \nu_1 = 2*3{,}14*2{,}0*6{,}5 = 8{,}98м/сек \quad (2.2.4)$$

$$F_1 = ma - F_{раб} = m(a - \mu g) = m(\frac{\upsilon_1^2}{R} - \mu g) = 6*10^{-3}((\frac{8{,}98^2}{0{,}20}) - 0{,}25*9{,}8) = 2{,}2N \quad (2.2.5)$$

Here μ=0.25- is the coefficient of friction of cotton on the steel surface; *m* is the mass of the cotton fly, *g = 9.8m / sec* acceleration of free fall, *a- acceleration of the cotton* fly. According to the calculations made after the first stake drum on the cotton bale affects the force $F_1 = 2{,}2N$, with which the bale hits the mesh wall of the second stake drum with a linear velocity o_1 = *6.5m / sec.* When the cotton bale falls under its own weight onto the second snare drum, the following force acts on it from the side of this snare drum:

$$\nu_2 = 400об/мин = 6{,}67об/сек \quad (2.2.6)$$

$$\upsilon_2 = 2\pi R \nu_2 = 2*3{,}14*2{,}2*6{,}67 = 9{,}21м/сек \quad (2.2.7)$$

$$F_2 = m(\frac{\upsilon_2^2}{R} - \mu g) = 6*10^{-3}((\frac{9{,}21^2}{0{,}20}) - 0{,}25*9{,}8) = 2{,}3N \quad (2.2.8)$$

When a fly of cotton falls by its own weight onto the third snare drum, the following force acts on it from the side of this snare drum:

$$\nu_3 = 410об/мин = 6{,}83об/сек \quad (2.2.9)$$

$$\upsilon_3 = 2\pi R \nu_3 = 2*3{,}14*2{,}2*6{,}83 = 9{,}44м/сек \quad (2.2.10)$$

$$F_3 = m(\frac{\upsilon_3^2}{R} - \mu g) = 6*10^{-3}((\frac{9{,}44^2}{0{,}20}) - 0{,}25*9{,}8) = 2{,}4N \quad (2.2.11)$$

When a fly of cotton falls by its own weight onto the fourth snare drum, the following force acts on it from the side of this snare drum:

$$\nu_4 = 420об/мин = 7{,}0об/сек \quad (2.2.11)$$

$$\upsilon_4 = 2\pi R \nu_4 = 2*3{,}14*2{,}2*7{,}0 = 9{,}67м/сек \quad (2.2.12)$$

$$F_4 = m(\frac{\upsilon_4^2}{R} - \mu g) = 6*10^{-3}((\frac{9{,}67}{0.20}) - 0{,}25*9{,}8) = 2{,}5N \quad (2.2.13)$$ [7]

The obtained results have revealed that at vertical technology of cotton cleaning from small weed impurities the value of impact impacts is much smaller than in the horizontal technology of cleaning and practically does not affect the natural qualitative indicators of processed cotton.

It is determined that at horizontal technology of cotton cleaning from small weed impurities the coefficient of efficiency of "live" section does not exceed the value g|=(),25 : 0,30, while at vertical technology of cleaning this index is ^= 0,58^0,60, which is the main reason for improvement of cotton cleaning from small weed impurities.

In the set of the vertical cotton cleaner from fine weed impurities in the number of four stake drums, the acting forces on the cotton fly and the rotation frequency of the stake drums increase in the following sequence $F_1 < F_2 < F_3 < F_4$ and $v_1 < v_2 < v_3 < v_4$, which is the reason for the highly efficient vertical technology of cotton cleaning from fine weed impurities.

When cleaning cotton from fine weeds by the existing horizontal technology of cleaning, the rotation speed of staking drums is 9 m/sec.

At vertical technology of cotton cleaning there is a possibility of increasing the speed of rotation of staking drums up to the value of 9.67 m/sec. This is due to the fact that there are no counter impact impacts of staking drums, present in the horizontal technology of cotton cleaning, which negatively affect the natural quality indicators of processed cotton.

In connection with the absence of counter shock effects on the processed cotton, the loads on electric motors were considerably reduced, as a result of which instead of the total power at horizontal technology of cotton cleaning from fine weeds W= 11 kW*hour, at vertical technology of cleaning from fine weeds was W= 6 kW*hour.

2.3 Modelling the separation of fine weed impurities from the of cotton flow in vertical cleaning

As a result of their research, the authors developed a vertical cotton cleaner from small weed impurities [37 - p. 81-86]. In the section of cleaning from small weed impurities, staking drums are located parallel and vertical to the horizontal axis, while the cleaned cotton moves sequentially from one drum to another. The mesh surface is made in such a way that it embraces the staking drum at an angle of more than 1800 and the number of stakes of the drum is much more involved in the cleaning process due to which the friction coefficient of the cleaned cotton on the mesh surface is significantly increased and, as a consequence, an increase in the cleaning effect is achieved. On the existing and operating fine litter cleaners the angle of girth of the staking drum with the mesh surface does not exceed 1000 (Fig.2.10).

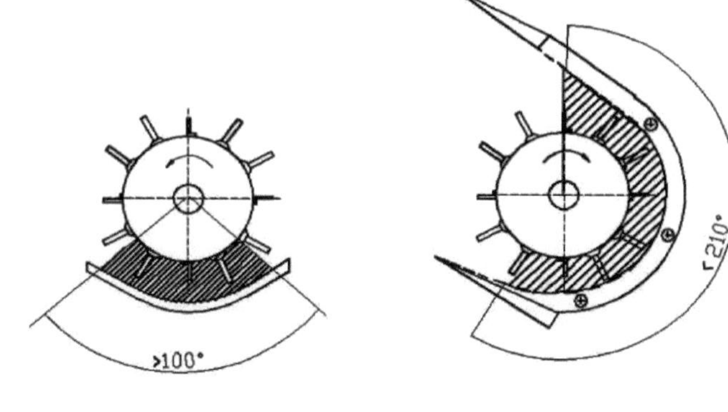

a) б)

Fig.2.10 Existing (a) and proposed (b) main working bodies of fine scum cleaners

To study the influence of the friction coefficient of the main working bodies on the efficiency of cotton cleaning from fine weed impurities with the help of application programmes, we carried out the following theoretical studies.

We assume that the cotton flow moves along the mesh surface along the arc AB and as a result of the impact of the stakes of the staking drum, fine weed impurities are extracted from it.

As a result of the rotational motion from the first stake drum, a stream of cotton is fed to the second stake drum along the arc BC.

In this interval, the separation of weed impurities is not carried out and the density of the cotton flow changes. Under the impact of the second spike drum and changing the position of the moving flow, there is a separation of weed impurities from the reverse side (surface) of the flow, that is, when the flow of cotton with the help of the first spike drum on the grid surface there is a separation of fewer weeds, as the cotton stream approaches the first drum in a less swollen state than the second drum, when the cotton stream passes through the second drum there is a more intensive release of weed impurities, as the cotton stream is cleaned in a more swollen state and the back side of the cotton stream is cleaned as well. The scheme of stationary state of this process is shown in Fig. 2.11.

The cotton flow parameters such as flow velocity, density and pressure respectively denote by v, p and p, the number of stakes of the staking drum denote by - n.

When the mesh surface, the pegs of the peg drum and the cotton interact, a pressure p acts on the cotton from the side of the pegs.

In this case, for the cotton flow element $s=Ra$ moving along the mesh surface, we can present the Euler equation in the following form:

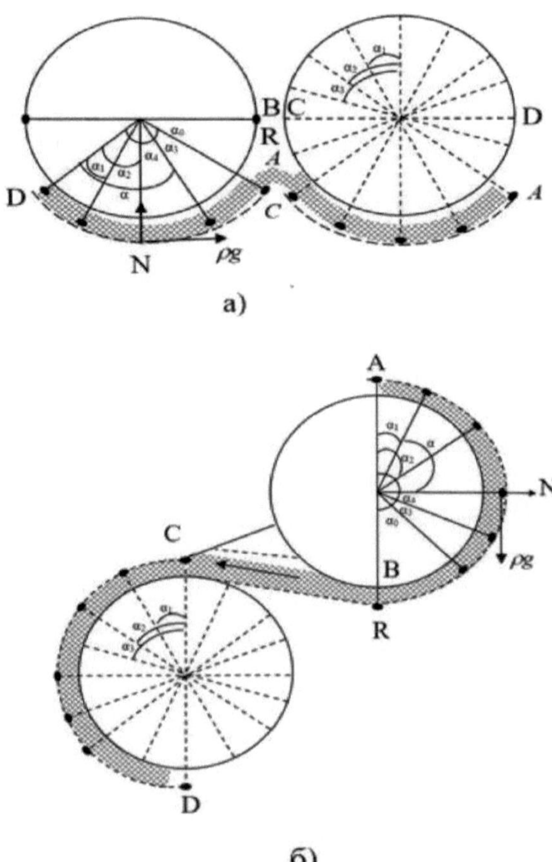

Fig.2.11 Schematic of cotton movement along the mesh surface and the cotton stake drum in the existing (a) and proposed (b) designs.

$$v\rho\frac{dv}{d\alpha}=-\frac{dp}{d\alpha}+\rho gR\sin\alpha-RNf \qquad (2.4.1)$$

Here α is the pole angle, R- radius of the stake, $\alpha_i = s_i/R$- pole angle of the inscribed *i-th* stake *f*- coefficient of friction between the mesh surface and the cotton flow, N- normal force generated by the mesh surface is determined by the following formula:

$$N=\rho\frac{v^2}{R}+pg\cos\alpha \qquad (2.4.2)$$

The value of the normal force (2.4.2) into the expression (2.4.1) we have:

$$v\rho\frac{dv}{d\varepsilon}=-\frac{dp}{d\alpha}+\rho gR(\sin\alpha+f\cos\alpha)-\rho v^2 f/R \qquad (2.4.3)$$

To complete equation (2.4.3), we take the following additional conditions.
1. There must be a relationship between pressure and density, i.e. equations reflecting this state are needed. In this work-dependent equation, the relationship between density and pressure in a stationary medium will be appropriately reflected as a linear relationship [48].

$$\rho=\rho_0[1+A(p-p_0] \qquad (2.4.4)$$

Here ρ_0, p_0 - is the state of density and pressure of cotton flow before entering the cleaning zone, A is an experimentally determined coefficient, which is the inverse of the bulk compression modulus of cotton K. It is assumed that equation (2.4.4) is appropriate for all zones of cotton flow. 2. The second condition is that the mass of the cotton flow is constant after a unit of time and this condition reflects the law of conservation of mass.

$$Q_0=\rho_0 v_0 h_0 L=\rho vhL \qquad (2.4.5)$$

Here, h0-thickness *of* cotton layer, ρ, v, h- density, velocity and thickness of cotton in an arbitrary arc of the cleaning zone. *L is the* length of the shaft.
In the calculations we assume that the cotton thickness h is a constant value.
Using formulae (2.4.4) and (2.4.5), *we express density and pressure through velocity v* at $A<<1$.

$$\rho=\frac{v_0\rho_0}{v}, \quad p=p_0-\frac{1}{A}\left(\frac{v}{v_0}-1\right) \qquad (2.4.6)$$

Substituting the expressions ρ and *p* into equation (2.3.3) we obtain the following equation for the cotton flow velocity:

$$\left(1-M^{-2}\right)\frac{dv}{dx}=\frac{gR}{v}(\sin\alpha+f\cos\alpha)-fv \qquad (2.4.7)$$

Here $M=\frac{v_0}{c_0}, c_0=\sqrt{\frac{1}{A\rho_0}}$ - shortness of wave distribution in cotton.

Multiplying equation (2.4.7) by v , with respect to the function $y=v^2(x)$ we obtain the following linear equation:

$$\frac{dy}{da}+2k_1 y=k_2(\sin\alpha+f\cos\alpha) \qquad (2.3.8)$$

,

here $k_1=2fM^2/(M^2-1), \; k_2=2gRM^2/(M^2-1)$

The angles between the grates are expressed through $\alpha=\alpha_{i-1}\ (i=1..n)$ and we assume that under the influence of the grates $\alpha=\alpha_{i-1}$ the flow gains velocity $v=v_k$ solution of the equation for each interval will have the form:

$$y_i=\exp[-k_1(\alpha-\alpha_{i-1})]\{v_k^2+k_2[F(\alpha)-F(\alpha_{i-1})]\}\quad \alpha_{i-1}<\alpha<\alpha_i \qquad (2.4.9)$$

$$F(\alpha)=\frac{g\,\mathrm{Re}^{K_1\alpha}}{1+K_1^2}[(K_1 f-1)\cos\alpha+(K_1+f)\sin\alpha]$$

Here n is the number of stakes,

The speed of the cotton according to formula (2.4.9) depends on the number $M=v_0/c_0$. In aerodynamics, this number is called the *Max number*.

When moving bodies in the air medium, their aerodynamic parameters depend on this number, the air resistance and aerodynamic forces affecting the movement of bodies depend on the value of the *Max* number. The occurrence of this number in the movement of cotton flow is explained by the linear relationship between density and pressure in a stationary medium (2.4.4). Under different conditions *M<1* and *M>1 the* flow velocity will be different. At values $M<1$ there is an acceleration in the movement of the medium, so the speed increases according to the formula (2.4.8) and the density decreases, that is, the medium is additionally loosened. Let us carry out a theoretical analysis of the process of separation of weed impurities based on the model of A.G.Sevostyanov [20 - p.345-349].

The amount of extracted weed impurities based on the formula for determining the flow rate (2.4.9).

We assume that the number of stakes of the staking drum is six units and they are installed at appropriate angles $\alpha_0=0$, $\alpha_1=30^0$, $\alpha_2=60^0$, $\alpha_3=90^0$, $\alpha_4=120^0$, $\alpha_5=150^0$, $\alpha_6=180^0$. Determine the number of separated weed impurities in the first section in steady state. According to the adopted model, as a result of separation of weed impurities in each section, the differential mass of cotton flow dm will be determined according to the following regularity.

$$\frac{dm}{m}=\lambda\frac{d\rho}{\rho} \qquad (2.4.10)$$

Here λ is the experimental coefficient. Equation (2.4.10) is integrated by the condition $\rho(\alpha_0)=\rho_0$. The experimental coefficient is a coefficient that reflects the increase in the amount of emitted weed impurities with the increase in the area of the working area of the working zone of the cotton picker.

$$m = m_0(\rho / \rho_0)^{\lambda} \qquad (2.4.11)$$

Here MO is the mass of weed impurities that have not been cleaned in the cotton cleaning zone.

We use this formula for the first and second stakes, i.e. according to the formula denoting the reduced mass (2.4.11) through $\alpha_0 < \alpha < \alpha_1$ for the interval expression $\varepsilon_1 = 1 - \frac{m_1(\alpha)}{m_0}$ we take the cleaning effect of cotton,

m O

between the first two pegs of the snare drum.

The ratio of this difference $\Delta m_1 = m_0 - m_1(\alpha_1)$ would be

$$\Delta m_1 / m_0 = 1 - \varepsilon_1(\alpha_1) \qquad (2.4.12)$$

The relative amount of extracted weed impurities from the cotton stream during its cleaning between the first two stakes will be $(\alpha_0 < \alpha < \alpha_1)$. This indicator between the second and third stakes will be equal to $(\alpha_1 < \alpha < \alpha_2)$ and the cleaning effect is determined, according to equation (2.4.11), by the following formula $\varepsilon_2 = \varepsilon_1(\alpha_1)\rho(\alpha)/\rho(\alpha_1)$.

Using this method, we determine the amount of weed impurities extracted during the cleaning of the cotton stream between successive sections of barks:

$$\Delta m_2 / m_0 = (m_0 - \Delta m_1)\varepsilon_2(\alpha_2)/m_0 = [1 - \varepsilon_1(\alpha_1)]\varepsilon_2(\alpha_2) \qquad (2.4.13)$$

$$\Delta m_3 / m_0 = [1 - \varepsilon(\alpha_1)][1 - \varepsilon(\alpha_2)]\varepsilon_3(\alpha_3)$$

$$\Delta m_4 / m_0 = [1 - \varepsilon_1(\alpha_1)][1 - \varepsilon_2(\alpha_2)][(1 - \varepsilon_3(\alpha_3)]\varepsilon_4(\alpha_4) \qquad (2.4.14)$$

$$\Delta m_5 / m_0 = [1 - \varepsilon_1(\alpha_1)][1 - \varepsilon_2(\alpha_2)][(1 - \varepsilon_3(\alpha_3)](1 - \varepsilon_4(\varepsilon_4)]\varepsilon_5(\alpha_5)$$

$$\varepsilon_i = \varepsilon_{i-1}(\alpha_{i-1})\rho(\alpha)/\rho(\alpha_{i-1})$$

The graphs of distribution of cleaning efficiency coefficient ε_i from friction coefficient f and ratio M = Voo / s 0 at different values of distribution along the cotton cleaning arc are given. When carrying out calculations the values are taken $R = 0.2 м$, $v_k = 9 м/c$, $\lambda = 0.5$.

Fig. 2.12 and 2.13 show the graphs of change of the maximum value of the cleaning efficiency coefficient <<:at two values of the *friction* coefficient f *in* relation to the number M. From the analysis of graphs it is obvious that at small values of number M intensive separation of weed impurities takes place in sections between the first and the second stake of a stake drum, at increasing of number *M the* process of cleaning takes place in all points of section and intensity of cleaning gradually decreases. The increase of friction coefficient is observed in the first zone between the grate stakes, which is the reason for the increase of cleaning effect coefficient.

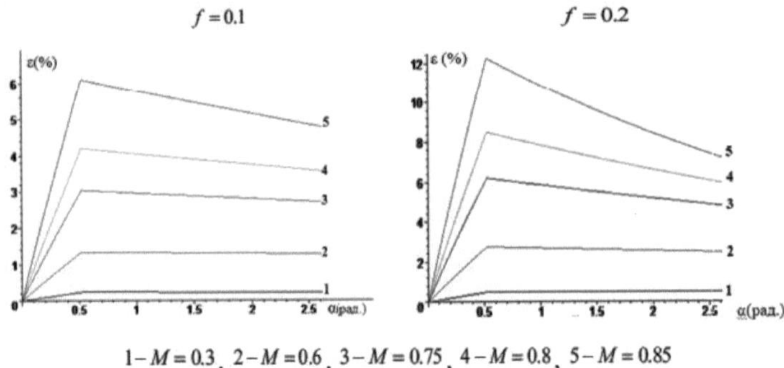

$1-M=0.3$, $2-M=0.6$, $3-M=0.75$, $4-M=0.8$, $5-M=0.85$

Fig. 2.12 Plots of distribution along the cleaning arc of change of cleaning efficiency coefficient$^{\varepsilon}$ at two values of *friction* coefficient *f* and different valuesk of number *M*

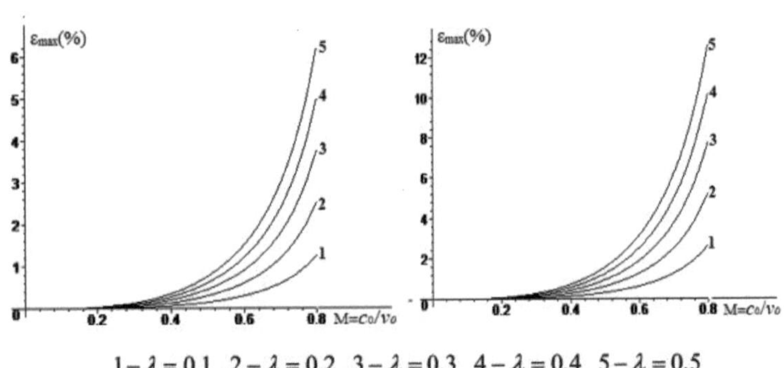

$1-\lambda=0.1$, $2-\lambda=0.2$, $3-\lambda=0.3$, $4-\lambda=0.4$, $5-\lambda=0.5$

Fig.2.13 Graph of variation of the maximum value of the cleaning efficiency coefficient ε *at* two values of the *friction* coefficient *f in* relation to the number *M*

From analysing the graphs, it can be concluded that as the number *M* approaches the value of 1, there is a sharp increase in the purification effect coefficient.

The process of separation of weed impurities from cotton can be obtained using formula (2.4.11). The process of intensity of picking out of weed impurities depends on the mutual location of the fine picking drums and on the method of transfer (transport) of cotton between them. In this connection, let us theoretically consider the schemes of two cleaning technologies, horizontal and vertical (proposed), shown in Fig. 2.4.2 a) and b).

The horizontal (existing) technology of cotton cleaning from fine weeds used in production has a number of existing disadvantages. One of them is the low cleaning effect, which is 40° - 45° . When moving horizontally while transferring cotton from one working body to another, there is also a loss of cotton velocity, which negatively affects the natural quality parameters of the cleaned cotton. Another disadvantage of this technology is that the transported cotton during cleaning from small weed impurities is mainly cleaned only from one side, which is in contact with the mesh surface, which affects the cleaning effect of the machine. As can be seen from Fig. 2.4.2-(b) the cleaning sections are arranged vertically relative to each other and in the boundary zones of the cotton transfer the speed remains unchanged, or in some cases increases.

In order to determine the cleaning effect of the modernised vertical cleaner, we will carry out a theoretical analysis of the movement and cleaning of cotton, according to the diagram in Fig. 2.4.2 a) and b). As can be seen from the scheme 2.4.2 a) in the existing operating cotton cleaner of horizontal layout, in the working area of the cleaning staking drum, in the process of cleaning cotton involves four stakes, and in the working area of the modernised cotton cleaner with a vertical layout of working bodies in the process of cleaning cotton on the staking drum involved 6 stakes. In the stationary mode of cotton cleaning in the horizontal arrangement when filling all four sections of the staking drum occurs process when filling the first section of the staking drum with cotton leads to the flight of partially cleaned cotton from the fourth section. If we assume that the angular velocity of the stakes is equal to ω . the time of periodicity of repetition of the cleaning process to take t, while if we neglect the time of movement from one cleaning section to another, the time required to clean cotton in all four cleaning sections will be equal to $4\pi/\omega$. With the help of the coefficients of cleaning efficiency, in the first approximation, determine the amount of weed impurities that are released from the cleaning sections. The relative amount of weed impurities released in all sections is defined as the following sum in the first approximation. For the time interval $T = 4\pi/\omega$ the total value of extracted impurities will be as follows

$$M_i = M_{i1} + M_{i2} + M_{i3} + M_{i4} \quad (i = 1,2,3,4) \qquad (15)$$

$$M_i = M_{i1} + M_{i2} + M_{i3} + M_{i4} + M_{i5} \quad (i = 1,2,3,4,5) \qquad (16)$$

Here the value of the numbers M_{ij} is the total value of the mass of weed impurities extracted in section i between stakes j and j +1 (j = 1...4). The formula for calculating these values depends on the design and method of cotton cleaning. In the first approximation, the numerical values of M_{ij} in different cleaning schemes, according to the formulas (2.4.12-2.4.14) is determined by the formulas.

$$M_{11}=d\bar{m}_{11}=\varepsilon_1(\alpha_1),\ M_{12}=(1-d\bar{m}_{11})\varepsilon_2(\alpha_2),\ M_{13}=(1-d\bar{m}_{11}-d\bar{m}_{12})\varepsilon_3(\alpha_3),$$

$$M_{14}=(1-d\bar{m}_{11}-d\bar{m}_{12}-d\bar{m}_{13})\varepsilon_4(\alpha_4),\ M_{15}=(1-d\bar{m}_{11}-d\bar{m}_{12}-d\bar{m}_{13}-d\bar{m}_{14})\varepsilon_5(\alpha_5) \quad (2.4.17)$$

$$M_{21}=(1-d\bar{m}_{11}-d\bar{m}_{12}-d\bar{m}_{13}-d\bar{m}_{14}-d\bar{m}_{15})\varepsilon_6(\alpha_1),\ M_{22}=(1-d\bar{m}_{21})\varepsilon_7(\alpha_2),$$

$$M_{23}=(1-d\bar{m}_{21}-d\bar{m}_{22})\varepsilon_8(\alpha_3),M_{24}=(1-d\bar{m}_{21}-d\bar{m}_{22}-d\bar{m}_{23})\varepsilon_9(\alpha_4),$$

$$M_{25}=(1-d\bar{m}_{21}-d\bar{m}_{22}-d\bar{m}_{23}-d\bar{m}_{24})\varepsilon_{10}(\alpha_5) \quad (2.4.18)$$

$$M_{31}=(1-d\bar{m}_{21}-d\bar{m}_{22}-d\bar{m}_{23}-d\bar{m}_{24}-d\bar{m}_{25})\varepsilon_{11}(\alpha_1),\ M_{32}=(1-d\bar{m}_{31})\varepsilon_{12}(\alpha_2),$$

$$M_{33}=(1-d\bar{m}_{31}-d\bar{m}_{32})\varepsilon_{13}(\alpha_3),\ M_{34}=(1-d\bar{m}_{31}-d\bar{m}_{32}-d\bar{m}_{33})\varepsilon_{14}(\alpha_4)$$

$$M_{35}=(1-d\bar{m}_{31}-d\bar{m}_{32}-d\bar{m}_{33}-d\bar{m}_{34})\varepsilon_{15}(\alpha_5) \quad (2.4.19)$$

$$M_{41}=(1-d\bar{m}_{31}-d\bar{m}_{32}-d\bar{m}_{33}-d\bar{m}_{34}-d\bar{m}_{35})\varepsilon_{16}(\alpha_1),M_{42}=(1-d\bar{m}_{41})\varepsilon_{17}(\alpha_2)$$

$$M_{43}=(1-d\bar{m}_{41}-d\bar{m}_{42})\varepsilon_{18}(\alpha_3),\ M_{34}=(1-d\bar{m}_{41}-d\bar{m}_{42})\varepsilon_{19}(\alpha_3)$$

$$M_{45}=(1-d\bar{m}_{41}-d\bar{m}_{42}-d\bar{m}_{43}-d\bar{m}_{44})\varepsilon_{20}(\alpha_5) \quad (2.4.20)$$

The full values of $\bar{m}_{ij}=m_{ij}/m_0$ m_{ij} Ba ε_i wa *£i are* given in the appendices.

In the existing designs of fine sap cleaners of horizontal layout, the amount of extracted weeds in one cleaning section is calculated according to the formula $M=\sum_{i=1}^{4}\sum_{j=1}^{4}M_{ij}$, and the amount of extracted sap in the

The cleaning section of the proposed vertical fine sod cleaner is calculated according to the formula $M=\sum_{i=1}^{4}\sum_{j=1}^{5}M_{ij}$.

The following values of the parameters $v_k=9\text{м/с},\ R=0.2\text{м},\ \lambda=0.25$. were assumed in the calculations The calculations were carried out at four values of the number of

M. As it was noted above, at horizontal arrangement of cleaning sections of the fine sap cleaner at transfer from one cleaning section to another cotton flow velocity is partially reduced, so this indicator is taken into account in the calculations.

Therefore, for each fines removal section, the corresponding values of the number *M* (Table 2.1-2.2) are adopted in the calculations.

Cleaning effect in the existing and proposed technology of cotton cleaning from fine litter at *friction* coefficient *f=0.1*

Table 2.1.

Values of the number *M*	f=0.1									
	Horizontal cleaning technology (existing), No. of sections and cleaning percentage					Vertical cleaning technology (proposed), No. of sections and cleaning percentage				
	1	2	3	4	**Total**	1	2	3	4	**Total**
0,3	0,47	0,34	0,29	0,24	**1.34**	0,59	0,59	0,43	0,43	**2.03**
0,38	0,90	0,79	0,69	0,60	**2.99**	1,13	1,13	0,98	0,97	**4.20**
0,56	2,64	2,28	2,01	1,77	**8.70**	3,28	3,28	2,76	2,67	**11.98**
0,75	11,36	9,99	7,29	4,42	**33.06**	13,79	13,79	8,72	7,56	**43.86**

Cleaning effect in the existing and proposed technology of cotton cleaning from fine litter at *friction* coefficient *f=0.2*

Table 2.2.

Values of the number *M*	f=0.2									
	Horizontal cleaning technology (existing), No. of sections and cleaning effect					Vertical cleaning technology (proposed), No. of sections and cleaning effect				
	1	2	3	4	**Total**	1	2	3	4	**Total**
0,3	0,97	0,70	0,59	0,49	**2.75**	1,21	1,21	0,87	0,86	**4.15**
0,38	1,86	1,60	1,39	1,20	**6.04**	2,31	2,31	1,95	1,91	**8.47**
0,56	5,32	4,41	3,76	3,25	**16.74**	6,57	6,57	5,11	4,78	**23.02**
0,75	21,09	15,27	9,31	6,33	**51.99**	24,91	24,91	11,57	8,48	**69.87**

III-CHAPTER

RESEARCH ON DEVELOPMENT OF A VERTICAL COTTON CLEANER FROM FINE LITTER

3.1 Analysing the results of laboratory tests of a vertical cotton cleaner from fine weed impurities

On the basis of the approved programme and methodology of tests (Appendix) comparative experimental studies of laboratory installations of horizontal arrangement (1HK) were carried out on the basis of the scientific laboratory of the department "Technology of primary processing of natural fibres", and the cotton cleaner from fine sorghum of vertical arrangement were carried out in the experimental-laboratory shop at the scientific centre of JSC "Paxtasanoat ilmiy markazi", where the length of the working part of the cone drum was made in the size of 300 mm.

The rotating parts were mounted on cantilever shafts. Grid surfaces and other elements were fixed on the body of the laboratory unit. The number of revolutions of cone drums for cleaning from fine weed impurities after feeders was changed according to the following increasing sequence: 1 - drum - 390 r/min, 2 - drum - 400 r/min, 3 - drum - 410 r/min and 4 - drum - 420 r/min, which allowed uniformly and continuously transport the cotton along the cleaning line without slaughtering.

For visual observation of the cotton cleaning process, the front part of the laboratory unit is covered with transparent organic glass, which allows video recording of the cotton cleaning process. When creating the experimental unit, the arc of girth of the mesh surface of the cone drum reached a maximum value of 210°.

Laboratory installation of vertical cleaner (Fig.3.1 and 3.2) has the following dimensions width - 1200 mm, height 2000 mm, the distance between the feed rollers 250 mm, the distance between the feed rollers and the first stake drum is 340 mm, the inter-axial distance between the stake-planate drums is 350 mm, the distance between the fourth stake-planate drum and weed screw is taken 310 mm, the length of the working part of all units of the laboratory installation is 300 mm.

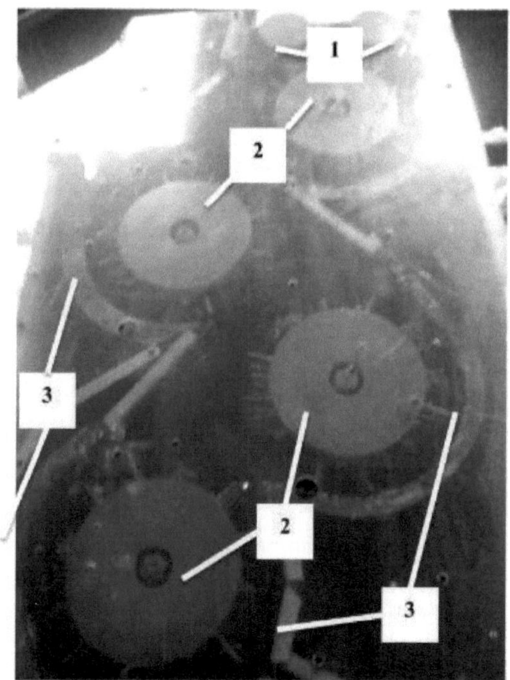

Fig.3.1 Laboratory installation of the modernised fine sod cleaner with vertically and parallel sequentially arranged stake and plate drums

1- feed rollers; 2- stake-plate drums; 3- satchel surface

For visual observation of the laboratory tests of the vertical cleaner, the front of the machine is covered with transparent glass. Due to the maximum circumference of the staking drums by the mesh surface, the maximum participation of the stakes in the process of cleaning the cotton from fine litter is achieved.

In the laboratory unit the cotton is fed by feed rollers 1 to the first cone drum 2, which cleans the cotton by dragging it over a mesh surface rotating at a speed of 390 rpm and throws it over. Cotton hitting the mesh surface 3 falls on the next second cone drum, which cleans the cotton dragging it over the mesh surface rotating at a speed of 400 rpm, while the flow of cleaned cotton changes the cleaning surface to the opposite plane. This process is repeated on the third and fourth stake and plate drums. The zigzag movement of the cotton to be cleaned and the maximum circumference of the mesh surface of the parallel perpendicularly arranged stake and slat drums allows to ensure a high cleaning effect of the machine [38 -p.44-50].

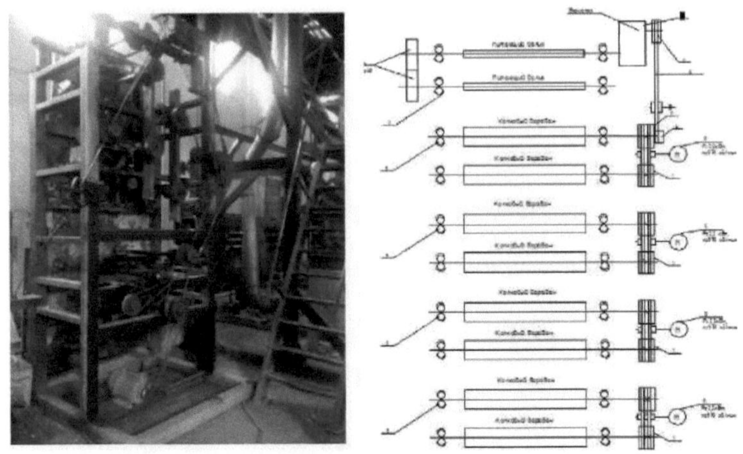

Fig.3.2 Schematic diagram and view of kinematic and working units of vertical combined cleaner of cotton from fine and coarse sap

Before the laboratory tests, according to the Test Programme, the productivity, number of revolutions of the working units and their technical condition, technological clearances between the mesh surface and stake drums, as well as other accepted control dimensions and condition of the laboratory unit units were checked. The diameter of two feed rollers is accepted d=150 mm, the diameter of four stake-plank drums D=400 mm, the gap between the stake-plank drum and the mesh surface is 14^18 mm. Laboratory tests were carried out under the following conditions:

- by means of stake drums in the existing cotton cleaning process;
- in the technological process of cotton cleaning by means of increased cotton contact area between the stake drum and the mesh surface;
- in the technological process, when the maximum degree of cotton loosening is ensured during the transfer of cotton from one picking drum to the next, due to the shockless trajectory of its movement;
- the possibility of cleaning the cotton from different sides (shredding) with the stakes and mesh surface of the cleaner when moving the cotton from one drum to another;
- the force of the cotton hitting the mesh surface, when high loosening is achieved, the fine weed impurities significantly lose their adhesion to the fibre, resulting in intensive cleaning of the cotton;
- during the technological process, when cone drums are arranged in a "zig-zag" pattern, i.e. in the vertical plane on one line and in an opposite-parallel manner.

During the experiments, the working zone of cotton cleaning from small weed impurities "stake drum - cotton - mesh surface" was varied in three ranges: minimum 1800, average 210 0, maximum 2400 (Fig.3.3) .

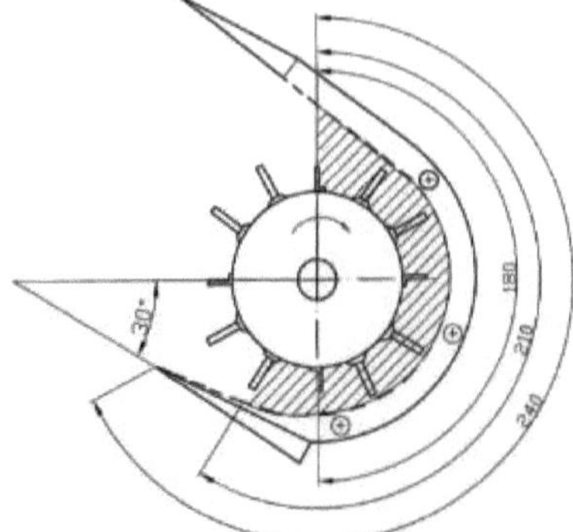

Fig. 3.3 Variants of the arc of coverage of the working area of cotton cleaning from fine weed impurities "Stake drum - cotton - mesh surface"

The linear speed of rotation of the stake-plate drums was 9 m/s as in the existing cotton cleaners from fine litter.

The samples obtained during the experiments were tested according to the methodological guidelines adopted in the national standards O'z DSt 643:2006, O'z DSt 644:2006, O'z DSt 592:2008 [39; 40; 41].

The moisture content of cotton was determined after weighing on electronic laboratory scales VLKT 500 determined on thermo-moisture meter VHS-M1. The weediness of samples was determined on the laboratory unit LKM and by the mass of extracted weed impurities the cleaning effect of the laboratory unit was deduced. The rotation speed and linear velocity of cone-plate drums were measured using a tachometer mark 8 TM 0.5 (accuracy class 2, unit of measurement 10 rpm).

The following process requirements apply to the modernised laboratory installation of the vertical cleaner:

- When cleaning cotton, to maximise the preservation of its natural quality characteristics;
- ensuring optimal performance of the laboratory unit at normal operation of all its components;
- when cleaning cotton to maximise loosening and prevent the density of the processed material from increasing;
- Ensuring complete removal of weed impurities while preventing the effect of their recirculation in the laboratory unit.

For carrying out comparative tests of technological indicators of laboratory units 1HK and vertical cleaner (Fig. 3.4), cotton C-6524, II - industrial grade, 3rd class with qualitative indicators: contamination 16.71% and moisture content 16.42% was used.

Fig. 3.4 Laboratory installations of the fine scum cleaner a) horizontal arrangement (1HK) and b) vertical arrangement

In the first stage of the tests, the efficiency of cleaning cotton from fine sap of the vertical cleaner was tested.

A comparative analysis of the cleaning effect of laboratory installations of cleaners with horizontal and vertical layout has shown (Fig. 3.5), that at vertical layout with initial high contamination of cotton 16, 71% cleaning effect is 49,67% at one-time cleaning of cotton, which is higher than the cleaning effect of horizontal layout by 5,44%. The cleaning effect is 45.62% in the vertical arrangement with an initial average cotton blockage of 5.48%, which is higher than the cleaning effect of the horizontal arrangement by 5.61%.

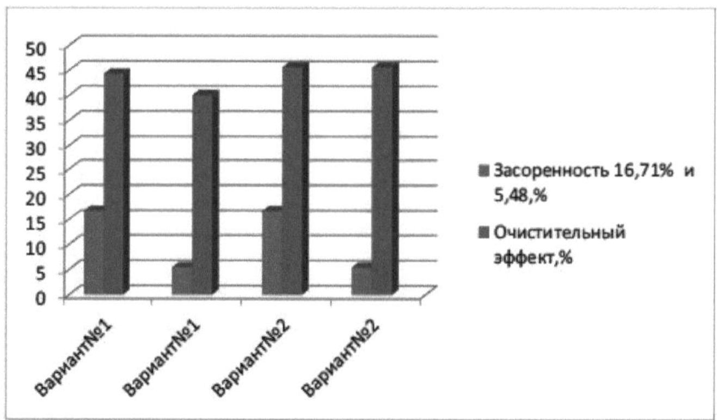

Fig.3.5 Comparative graph of cleaning effect of laboratory plants at 16, 71% and 5.48% clogging rate

Option #1 is to install a cleaner with a horizontal layout;

Option No. 2 - installation of a cleaner with a vertical layout.

When the effect of the laboratory plant capacity was studied, it was found that at a capacity of 7 t/h the cleaning effect was 40.1%, at a medium capacity of 5 t/h the cleaning effect was 41.4% and at a low capacity of 3 t/h the cleaning effect of the laboratory fine sorghum cleaner increased to 43.2% (Fig.3.6).

During the tests of the laboratory cleaner of vertical layout, the influence of cotton moisture content on the efficiency of its cleaning was studied. With increasing moisture content, a significant decrease in cleaning effect was observed, for example, when cleaning cotton with a high moisture content of 16.42% with a control blockage of 5.48%, after cleaning was 3.52% with a cleaning effect of 35.76%. In cleaning cotton with moisture content of 10.14% with control contamination of 5.48%, after cleaning the contamination was 3.38%. That is, the cleaning effect at this moisture content was higher and was equal to 38.32%. At the initial standard cotton moisture content of 8.52% and 5.48% clogging after cleaning, the clogging was 3.14%, and the cleaning effect of the laboratory installation of the vertical cleaner on fine sap was 42.70% (Fig. 3.7).

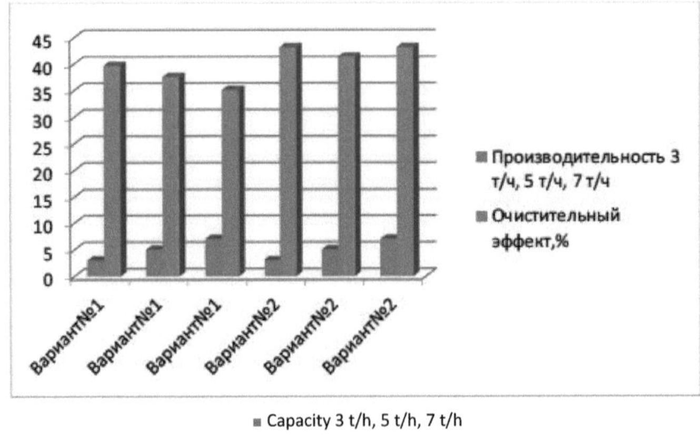

■ Capacity 3 t/h, 5 t/h, 7 t/h

■ Cleansing effect,%

Fig. 3.6 Comparative graph of the purification effect of laboratory plants at 3 t/h, 5 t/h and 7 t/h capacity

Option No. 1 - Horizontally mounted cleaner; Option No. 2 - Vertically mounted cleaner

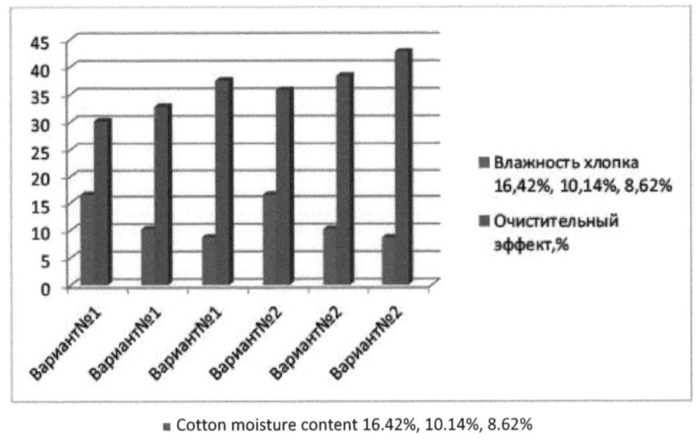

■ Cotton moisture content 16.42%, 10.14%, 8.62%

■ Cleansing effect,%

Fig. 3.7 Comparative graph of cleaning effect of laboratory plants at cotton moisture content of 16, 71%, 10.14% and 8.62%

Option No. 1 - Horizontally mounted cleaner; Option No. 2 - Vertically mounted cleaner

Test results of the vertical cleaner at the experimental laboratory workshop of Paxtasanoat ilmiy markazi S.A.

Table No. 3.1

№	Indicators	%	Cleansing effect, %
Influence of cotton contamination			
1	Control high infestation	16,71	
2	Debris after cleaning	8,41	49,67
3	Control average weediness	5,48	
4	Debris after cleaning	2,98	45,62
Influence of plant performance			
1	Control sample	5,48	
2	High capacity, 7 tonnes per hour	3,49	40,1
3	Average capacity, 5 t/h	3.21	41,4
4	Low capacity, 3 t/h	3,11	43,2
Influence of cotton moisture content			
1	At maximum humidity	16,42	
	Control weediness	5,48	
	Debris after cleaning	3,52	35,76
2	At average humidity	10,14	
	Control infestation	5,48	
	Debris after cleaning	3,38	38,32
3	At minimum humidity	8,52	
	Control infestation	5,48	
	Debris after cleaning	3,14	42,70

Analyses of comparative performance of two arrangements of fine sap cleaners showed high efficiency of the vertical cleaner (Tables 3.1 and 3.2).

Results of tests of the horizontal cleaner at the laboratory of the Department "Technology of primary processing of natural fibres"

Table No. 3.2

№	Indicators	%	Cleansing effect, %
Influence of cotton contamination			
1	Control high infestation	16,71	
2	Debris after cleaning	9,32	44,23
3	Control average weediness	5,48	
4	Debris after cleaning	3,29	40,01
Influence of plant capacity			
1	Control sample	5,48	
2	High capacity, 7 tonnes per hour	3,56	35,08
3	Average capacity, 5 t/h	3.42	37,55
4	Low capacity, 3 t/h	3,31	39,67

Influence of cotton moisture content			
1	At maximum humidity	16,42	
	Control infestation	5,48	
	Debris after cleaning	3,83	30,15
2	At average humidity	10,14	
	Control infestation	5,48	
	Debris after cleaning	3,68	32,78
3	At minimum humidity	8,52	
	Control infestation	5,48	
	Debris after cleaning	3,43	37,49

For example, when cleaning cotton from small weed impurities on a cleaner with a vertical layout, such indicators as cotton clogging, different values of its moisture content and cleaning effect at different values of plant capacity are significantly higher than similar indicators of tests of the plant with a horizontal layout. This is achieved by increasing the arc around the mesh surface of the stake drums, by increasing the impact of the cotton transfer and by changing the cleaning surface of the cotton flow as it is transported between the stake drums and cleaned in a reciprocal manner without the counter impact present in the existing horizontal small sap cleaners.

3.2 Planning experiments and analysing factors affecting the cleaning effect of the vertical cleaner

In order to investigate the influence on the cleaning effect on fine sap of the vertical cleaner of the working area of the mesh surface, cotton moisture content and productivity of the laboratory unit, the following main parameters are taken into account, which are accepted as input and output parameters:

x1 - working area of the mesh surface, a, degree.

It is known that the working area of the mesh surface in the cleaning of fine weed impurities significantly affects the efficiency of cleaning. As the working area increases, the separation of fine weed impurities improves. In existing fine weed cleaners, the girth angle of the mesh surface of the cone drums does not exceed the value of 100° . In the laboratory setup, we have taken the maximum value of the working area of the mesh surface as 2400 and the minimum value as 1800.

x_2 - cotton moisture content, *W*, %.

According to the requirements of the existing regulations, the standard moisture content of cotton for cleaning from fine weeds is 8-9%. It is known that a change in these normative values has a significant impact on the efficiency of cleaning. Therefore, we have adopted a maximum moisture content of 15 per

cent and a standard moisture content of 9 per cent.
x_3 - productivity, P, tonnes/hour.
The influence of productivity on the cleaning effect of fine sap cleaners has been proven by numerous studies. The operation of the laboratory cleaner with optimal productivity allows to ensure the preservation of natural quality indicators of processed cotton. Based on these positions, we have chosen the maximum capacity of the plant 7 tonnes/hour and the minimum 3 tonnes/hour.
In conducting the research a full-factor experiment was chosen [42 - 274 p.] 23 .
After selecting the main factors and levels of their variation, it was determined by which output parameters will be used to make a conclusion on the evaluation of the laboratory unit performance, as well as to optimise the technological and design parameters of the purifier. Based on the obtained parameters, a planning matrix was drawn up, which is presented in Table 3.3 and 3.4

Main factors and levels of variation

Table - 3.3.

№	Name of factor	Designation code	True values of the factor			Variation range
			-1	0	+1	
1	Working area of the mesh surface, a, degree.	x_1	180	210	240	30
2	Cotton moisture content, *W, %.*	x_2	9	12	15	3
3	Capacity, P, tonnes/hour.	x_3	3	5	7	2

When conducting experiments, the output parameter is chosen as the cleaning effect of the laboratory unit *(y)*. Further we build the plan of the experiment planning matrix, having written down in the standard form the conditions of experimentation in the form of a table, in the rows of which the data of experiments are written down in the columns "factors" in code designations, with realisation of all possible combinations of ordered combinations of factors, on the basis of which we conduct 2 parallel experiments.
Then the number of experiments is $N^{=2} \cdot {}^8$, taking into account the number of repetitions $m = ^2$ the total number of experiments is $N \cdot m = ^{16}$ The complete plan of the planning matrix is given in Table 3.2.7

Full outline of the planning matrix - 2^3

Table 3.4.

t/r	Factors		
	x_1, degree	x_2, %	x_z, tonnes per hour
1	180	9	3
2	240	9	3
3	180	15	3

4	240	15	3
5	180	9	7
6	240	9	7
7	180	15	7
8	240	15	7

The obtained experimental data were processed on the computer resulting in regression equations.

For this purpose, the factors influencing the maximum release of fine weed impurities smaller than 10 mm were studied. Experimental results are presented in Table 3.5.

Find the values of regression coefficients using the following formulas.

$$b_0 = \frac{1}{N}\sum_{u=1}^{N}\bar{y}_u,\ b_i = \frac{1}{N}\sum_{u=1}^{N}X_{iu}\bar{y}_u,\ b_{ij} = \frac{1}{N}\sum_{u=1}^{N}X_{iu}X_{ju}\bar{y}_u,\ b_{ijk} = \frac{1}{N}\sum_{u=1}^{N}X_{iu}X_{ju}X_{ku}\bar{y}_u \qquad (3.2.1)$$

Planning matrix, experimental and estimated indicators

Table - 3.5

No. of experience	Intermediate values of factors			Cleansing effect of the installation, $\bar{y}_r$		$\bar{y}_u$	S_u^2	Y_{Ru}	$R_0(\%)$
	X1	*X2*	*Hz*	$\bar{y}_{t1}$	$\bar{y}_{t2}$				
1	-	-	-	54,3333	55,0667	54,7000	0,2689	55,396	1,2561
2	+	-	-	60,4667	60,1333	60,3000	0,0556	60,192	- 0,1800
3	-	+	-	51,4667	51,1333	51,3000	0,0556	51,408	0,2107
4	+	+	-	56,4667	55,8000	56,1333	0,2222	55,438	- 1,2552
5	-	-	+	53,4333	52,0667	52,7500	0,9339	52,054	- 1,3367
6	+	-	+	54,4667	53,3333	53,9000	0,6422	54,008	0,2006
7	-	+	+	45,8333	44,8333	45,3333	0,5000	45,225	- 0,2395
8	+	+	+	51,7667	51,0333	51,4000	0,2689	52,096	1,3357

After determining the coefficients, let's write down the coded variable regression equations.

$$\hat{y} = b_0 + \sum_{i=1}^{k} b_i x_i + \sum_{i<1}^{k} b_{ij} X_i X_j + \sum_{i<j<l}^{k} b_{ijl} X_i X_j X_l \qquad (3.2.2)$$

To carry out calculations of coefficient values, we determine average values from the table.

average value

$$b_0 = \frac{1}{N}\left(\bar{y}_{cp_1} + \bar{y}_{cp2} + \bar{y}_{cp_3} + \bar{y}_{cp_4} + \bar{y}_{cp_5} + \bar{y}_{cp_6} + \bar{y}_{cp_7} + \bar{y}_{cp_8}\right) = 53{,}2271$$

Let us consider linear values of the coefficients

$$b_1 = \frac{1}{N}\left(x_{11}\overline{Y}_{cp_1} + x_{12}\overline{Y}_{cp2} + x_{13}\overline{Y}_{cp_3} + x_{14}\overline{Y}_{cp_4} + x_{15}\overline{Y}_{cp_5} + x_{16}\overline{Y}_{cp_6} + x_{17}\overline{Y}_{cp_7} + x_{18}\overline{Y}_{cp_8}\right) = 2,2063$$

$$b_2 = \frac{1}{N}\left(x_{21}\overline{Y}_{cp_1} + x_{22}\overline{Y}_{cp2} + x_{23}\overline{Y}_{cp_3} + x_{24}\overline{Y}_{cp_4} + x_{25}\overline{Y}_{cp_5} + x_{26}\overline{Y}_{cp_6} + x_{27}\overline{Y}_{cp_7} + x_{28}\overline{Y}_{cp_8}\right) = -2,1854$$

$$b_3 = \frac{1}{N}\left(x_{31}\overline{Y}_{cp_1} + x_{32}\overline{Y}_{cp2} + x_{33}\overline{Y}_{cp_3} + x_{34}\overline{Y}_{cp_4} + x_{35}\overline{Y}_{cp_5} + x_{36}\overline{Y}_{cp_6} + x_{37}\overline{Y}_{cp_7} + x_{38}\overline{Y}_{cp_8}\right) = -2,3812$$

Consider nonlinear values of the coefficients:

$$b_{12} = \frac{1}{N}\left(x_{11}x_{21}\overline{Y}_{cp_1} + x_{12}x_{22}\overline{Y}_{cp2} + x_{13}x_{23}\overline{Y}_{cp_3} + x_{14}x_{24}\overline{Y}_{cp_4} + x_{15}x_{25}\overline{Y}_{cp_5} + x_{16}x_{26}\overline{Y}_{cp_6} + x_{17}x_{27}\overline{Y}_{cp_7} + x_{18}x_{28}\overline{Y}_{cp_8}\right) = 0,5188$$

$$b_{23} = \frac{1}{N}\left(x_{21}x_{31}\overline{Y}_{cp_1} + x_{22}x_{32}\overline{Y}_{cp2} + x_{23}x_{33}\overline{Y}_{cp_3} + x_{24}x_{34}\overline{Y}_{cp_4} + x_{25}x_{35}\overline{Y}_{cp_5} + x_{26}x_{36}\overline{Y}_{cp_6} + x_{27}x_{37}\overline{Y}_{cp_7} + x_{28}x_{38}\overline{Y}_{cp_8}\right) = -0,2938$$

$$b_{123} = \frac{1}{N}\left(x_{11}x_{21}x_{31}\overline{Y}_{cp_1} + x_{12}x_{22}x_{32}\overline{Y}_{cp2} + x_{13}x_{23}x_{33}\overline{Y}_{cp_3} + x_{14}x_{24}x_{34}\overline{Y}_{cp_4} + x_{15}x_{25}x_{35}\overline{Y}_{cp_5} + x_{16}x_{26}x_{36}\overline{Y}_{cp_6} + x_{17}x_{27}x_{37}\overline{Y}_{cp_7} + x_{18}x_{28}x_{38}\overline{Y}_{cp_8}\right) = 0,7104$$

In this case, we adopt a multifactor model:

$$y_R = b_0 + b_1x_1 + b_2x_2 + b_3x_3 + b_{12}x_1x_2 + b_{13}x_1x_3 + b_{23}x_2x_3 + b_{123}x_1x_2x_3$$

To represent the model in its final form, we use the criterion
Student's t-test.

We check the significance of regression coefficients by Student's criterion.
Preliminarily, all regression coefficients are calculated at univariate

A b confidence range Δb.

$$S^2(\bar{y}) - \frac{1}{N}\sum_{u=1}^{N} S_u^2(y) = 0,0022153\,;$$

$$S(\bar{y}) = \sqrt{S(\bar{y})} = \sqrt{0,0022153} = 0,0166406$$

$$\Delta b = t_T \frac{S(\bar{y})}{\sqrt{N}} = 2,12 \cdot \frac{0,0166406}{\sqrt{8}} = 0,0497735$$

The tabular value of Student's coefficient is taken from the reference book:

$$t_T\left[P_D, f(S_u^2) = N(m-1)\right] = t_T\left[P_D = 0,95; f = 8\cdot(3-1) = 16\right] = 2,12.$$

Provided that when the regression coefficients are above the confidence coefficient, then they are considered significant.

From the obtained results we can see that the calculated values of the coefficients $b_0, b_1, b_2, b_3, b_{12}, b_{123} \geq \Delta b$ are higher than the tabulated values, so they are assumed to be all significant. According to these conditions, the coefficients b_{13} and b_{23} are not significant. As a result, we obtain the following model:

$$Y_R = 53,2271 + 2,2063x_1 - 2,1854x_2 - 2,3813x_3 + 0,5188x_1x_2 + 0,7104x_1x_2x_3 \quad (2.3.3)$$

To test the adequacy of the model, we use the Fisher's criterion formula. To do this, let us compare the experimental and calculated data of the output factor (Table 3.6):

Given the number of significant coefficients N_K=6, we have:

$$S^2_{\text{над}}(y) = \frac{\sum_{u=1}^{N}(\bar{y}_u - \bar{y}_{Ru})^2}{N-k-1} = 1{,}98$$

Table 3.6

u	$\bar{y}_u$	$\bar{y}_{Ru}$	$\bar{y}_u - y_{Ru}$	$(\bar{y}_u - y_{Ru})^2$
1	54,7	55,396	-0,696	0,484
2	60,3	60,192	0,108	0,012
3	51,3	51,408	-0,108	0,012
4	56,1	55,438	0,696	0,484
5	52,8	52,054	0,696	0,484
6	53,9	54,008	-0,108	0,012
7	45,3	45,225	0,108	0,012
8	51,4	52,096	-0,696	0,484
$\sum_{u=1}^{N}$	-	-	-	1,984

The obtained value is greater than the value $S^2(\bar{y}) = 0{,}3684$, so the calculated value of the criterion is determined by the formula:

$$F_R = \frac{S^2_{над}}{S^2_y} = 5{,}3845$$

Find the tabulated value of Fisher's coefficient:

$$F_T\left[P_D = 0{,}95; f(S^2_y) = 16,\ f(S^2_{\text{над}}) = 4\right] = 5{,}85.$$

Since from the condition, $F_R < F_T$, the model is considered adequate.

On the basis of the obtained results we obtain graphical dependences. The boundary values of the working area of the mesh surface, cotton moisture and productivity for cleaning cotton from fine sap with cleaning efficiency from 44% to 60% are determined.

The graphs (Fig.3.8) show the influence on the cleaning effect of different values of the angle of coverage of the working area of the mesh surface and the moisture content of cotton at a plant capacity of 3 tonnes/hour.

The analysis of the obtained graphical dependences shows that at the productivity of the cleaner in 3 tonnes/hour for achievement of the cleaning effect in 52% it is necessary the value of the angle of coverage of the working zone of the mesh surface in the interval $180^0<\alpha<195^0$ at humidity not exceeding the value from 12% to 15%.

With a cleaning capacity of 3 tonnes/hour, to achieve a cleaning effect of 56%, the angle of coverage of the working area of the mesh surface must be between $180^0<\alpha<240^0$ at a humidity not exceeding 13.5% to 15%.

To achieve the purification effect of fine weed impurities, 60% of the parameter values of the input factors are not satisfied.

Fig.3.9 shows the graphs of the influence on the cleaning effect of different values of the angle of coverage of the working area of the mesh surface and the moisture content of cotton at a capacity of 5 tonnes/hour.

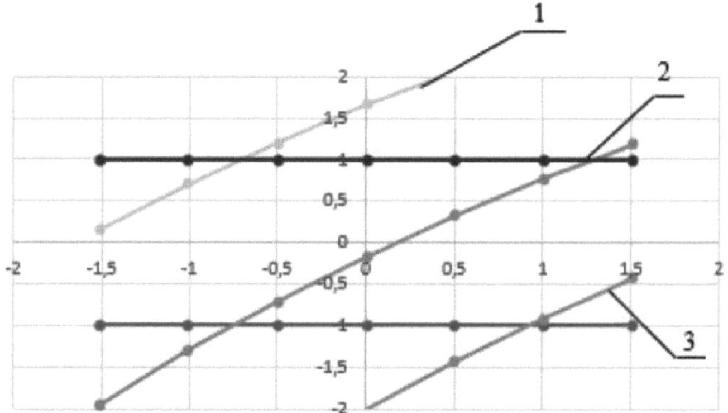

Fig. 3.8. Influence on the cleaning effect of different values of the angle of coverage of the working area of the mesh surface and the moisture content of cotton at a laboratory plant capacity of 3 tonnes/hour

1 - with a cleansing effect of 52%;
2- with a cleansing effect of 56%;
3- with a cleansing effect of 60%.

The analysis of the obtained graphical dependences shows that at productivity of the cleaner in 5 tonnes/hour for achievement of cleaning effect in 52% it is necessary value of angle of coverage of working zone of a grid surface in interval 1800<a<2250 at humidity not exceeding values from 10,5% to 15%. With a cleaning capacity of 5 tonnes/hour to achieve a cleaning effect of 56% it is necessary to have an angle of coverage of the working area of the mesh surface in the range 2250<a<2400 at a humidity not exceeding the value from 9% to 10.5%.

To achieve the purification effect of fine weed impurities, 60% values of input factor parameters are also not satisfactory

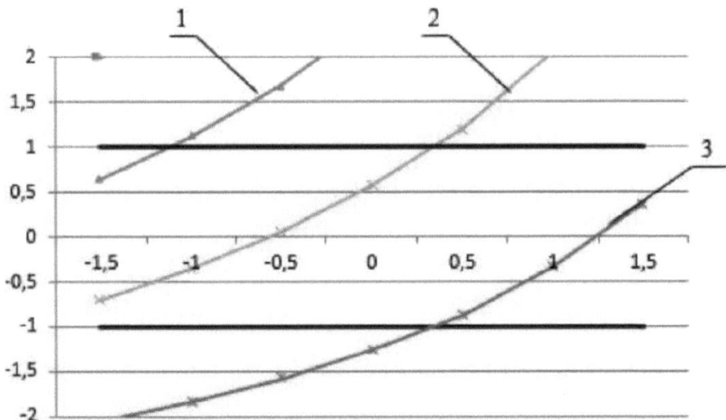

Fig.3.9 Influence on the cleaning effect of different values of the angle of coverage of the working area of the mesh surface and moisture content of cotton at the productivity of the laboratory plant 5 tonnes/hour

1- - with a cleansing effect of 48%;

2- with a cleansing effect of 52%;

3- with a cleansing effect of 56%. a

Fig.3.10 shows the graphs of the influence on the cleaning effect of different values of the angle of coverage of the working area of the mesh surface and the moisture content of cotton at a capacity of 7 tonnes/hour.

The analysis of the obtained graphical dependences shows that at productivity of the cleaner in 7 tonnes/hour for achievement of cleaning effect in 48 % it is necessary value of an angle of coverage of a working zone of a grid surface in interval 1800<a<2100 at humidity not exceeding value from 12 % to 15 %.

With a cleaning capacity of 5 tonnes/hour to achieve a cleaning effect of 52%, the angle of coverage of the working area of the mesh surface should be between 1800<a<2400 at a humidity not exceeding the value of 9% to 15%.

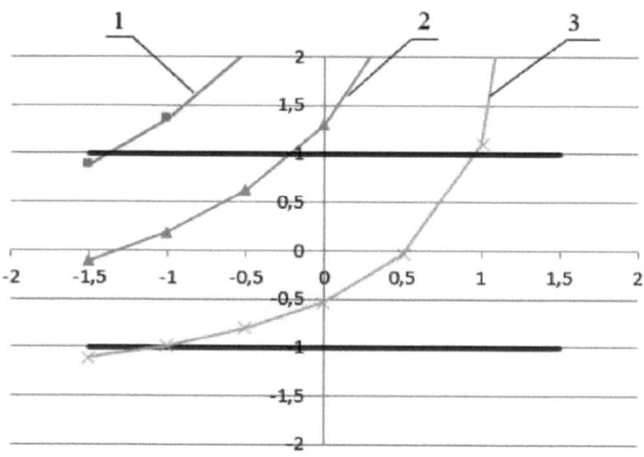

Fig.3.10 Influence on the cleaning effect of different values of the angle of coverage of the working area of the mesh surface and moisture content of cotton at the productivity of the laboratory plant 7 tonnes/hour

1- - with a cleansing effect of 44%;
2- with a cleansing effect of 48%;
3- with a cleansing effect of 52%.

On the basis of laboratory tests of the modernised cleaner with vertical arrangement of the fines cleaning section, planning of experiments and analysis of factors influencing the cleaning effect of the machine, it was found that the optimum cleaning effect of the machine is 56%.

3.3. Classification of cotton fines cleaners.

In order to identify and select research directions, based on the study and analysis of earlier scientific works in the field of cotton cleaning from fine weed impurities, there was a need for their classification, allowing to systematise and determine priorities in solving the set technological problems. Proceeding from this the previous classification schemes of cotton cleaning from fine weed impurities, developed by domestic scientists [15, p. 136] and [43, p.77], were studied. They introduce the classification of fine weed cleaners (Fig.3.11), which allows to reveal advantages and disadvantages of existing designs of cleaners, as well as their working units and principle of action. The analysis of these classification schemes allows to define the priority directions of research, showing significant reserves in increasing the efficiency of the process. Such directions are: rational use and increase of the path of cotton flow movement along the mesh surface (cleaning arc); selection of optimal modes of cleaning and routes of transit movement of cotton, at which its maximum cleaning is achieved; achievement of effective use of stakes and slat drums in the process of cotton cleaning from small weed impurities due to their spatial change of their ways of arrangement.

The basic for the proposed classification scheme of cotton cleaners from small weed impurities is the variant developed by A.E.Lugachev in the above-mentioned scientific work (Fig.3.12). The scheme is supplemented with a feature - the principle of cotton flow movement, which consists, in turn, of forced-impact and free-no-impact, smooth impact. Further, according to the direction of cotton flow the cleaners are divided into: horizontal, inclined, stepped and vertical, which are divided according to the angle of girth of the cone drum by the mesh surface into less than 90° and more than 90°. As in the module of cotton cleaning from small weed impurities transport-ripping organ along with the movement of cotton, is the cause of air flow, then when modernising the existing schemes of cotton cleaning, it is necessary to take into account and this factor of intensification of the cleaning process of cotton. Taking into account this classification scheme of cotton cleaners from small weed impurities the ways of their improvement are revealed.

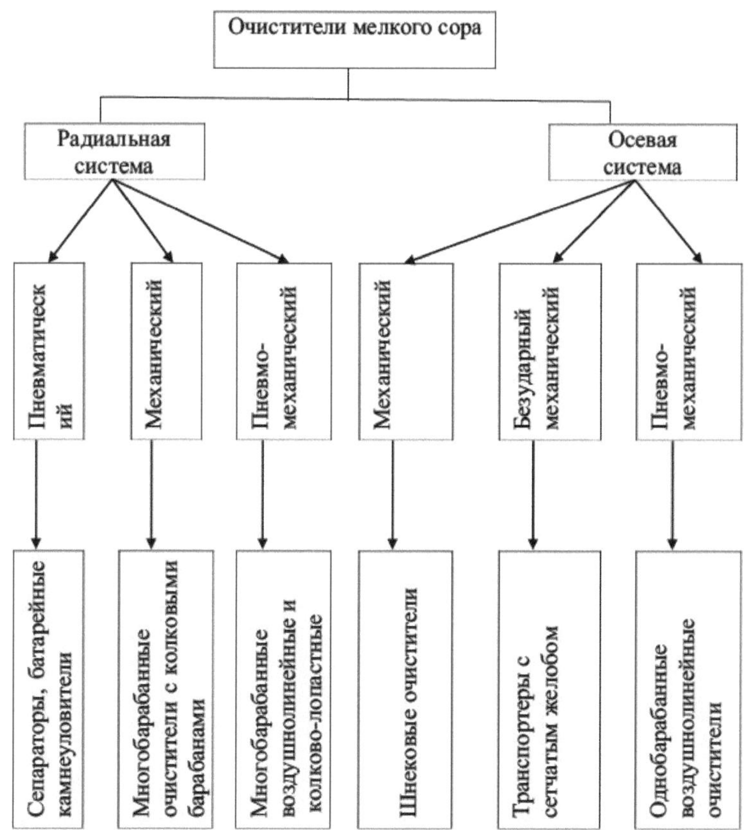

Fig.3.11 Classification of fine litter cleaners (by M.Jamalova)

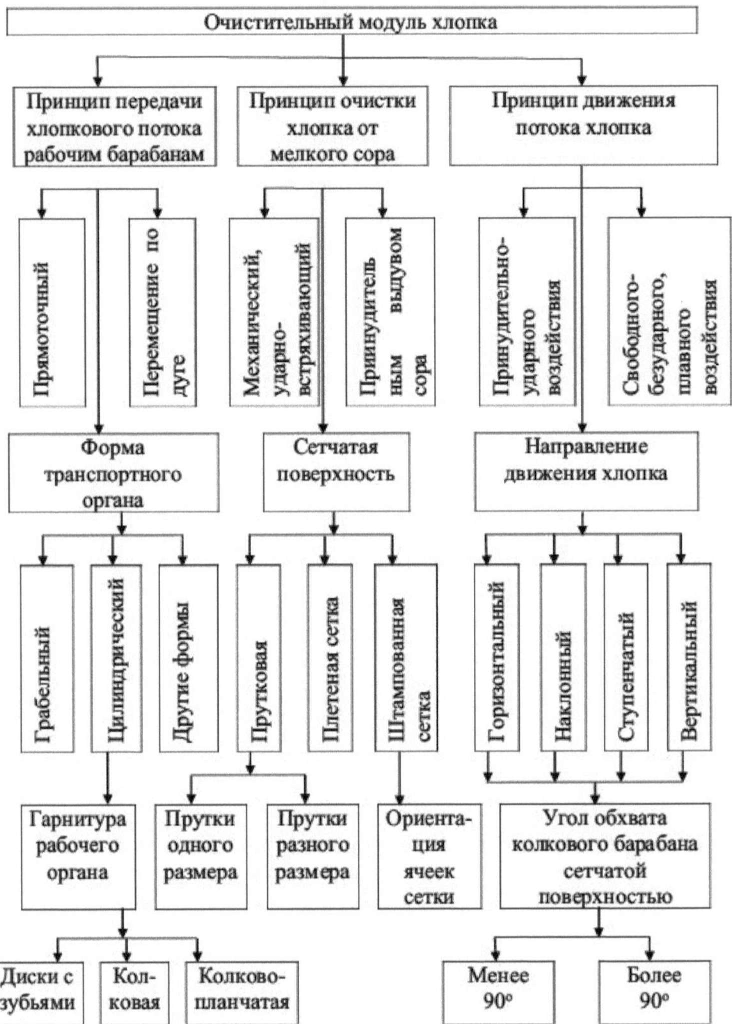

Fig.3.12 Classification scheme of cotton cleaners from small weed impurities.

3.4 Vertical fine debris cleaner

For cleaning of hand and machine picked cotton from small weed impurities with moisture content not exceeding 14%, cleaning machines of SCh-02, 1XK or 6A-12M1 brand are installed in cleaning shops.

Cleaners mark 1XK and СЧ-02 are also used as part of flow lines in cleaning and drying-cleaning shops of cotton plants with batteries of cleaners ЧХ-5, ЧХ-6 and ЧХ-3М2 with obligatory installation at the beginning of the technological

process of a catcher of heavy impurities.
Technological requirements for cleaners:
1.Ensuring maximum cleansing effect;
2. Avoidance of fibre and seed damage during cleaning;
3. minimising the waste of cotton bales with weed impurities.
In the 1HK fine cotton cleaners, the cotton transport process is carried out by the unidirectional movement of the cone drums. In the zone between two adjacent drums, which move towards each other, the cotton particles are subjected to significant impacts from the stakes of the adjacent drums. The linear velocity of the cone drums is V1 = 9 m/s, and the cotton particle is impacted at a velocity of V_2 = 18 m/s (the velocity of the adjacent cone drum is added). As a consequence, significant damage to fibre and seeds occurs [44 -p.27-32].
In continuation of the previously conducted research [15 - p. 111] we propose an improved scheme of vertical cotton cleaner from fine sap, which allows to eliminate the above drawbacks by sequential movement of stake-plank drums, where there is an opportunity to increase the angle of coverage of the mesh surface of the drum. The proposed scheme is depicted in Fig.3.13, which shows a cross section of it. The unidirectional speed of rotation of the cone drums allows to eliminate slaughtering situations in the machine. For this layout of the fine sap cleaner design, application No. FAP 20170134 dated 27 November 2017 was filed with the Intellectual Property Agency of the Republic of Uzbekistan. At the same time, the angle of girth of the cone drum by the mesh surface is more than 180° . The given scheme of arrangement of cleaning sections and consecutive transport of cotton on conjugated cleaning sections allows to increase significantly the cleaning effect, and also preserving natural qualitative indicators of cotton and its components, will save from damage fibres and seeds during transport on drums, that is the basis for development of technology of vertical cleaning of cotton on cotton mills.

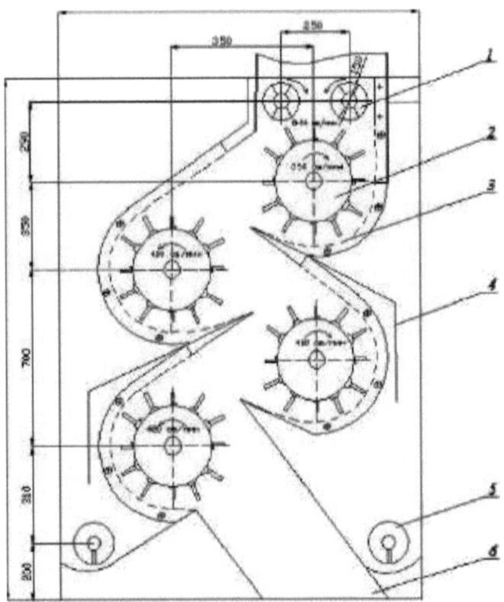

Fig. 3.13 Schematic diagram of a fine litter cleaner with vertically and parallel arranged stake and plate drums with sequential cotton movement

1 - feeder; 2 - feeder-plate drum; 3 - mesh surface;
4 - protective screen; 5 - picking auger; 6 - outlet pipe

The vertical fine picking cleaner with vertically and parallel located stake and slat drums with sequential movement of cotton includes a feeder 1, under which in the vertical plane, in the course of the process, the section of cotton cleaning from fine picking with stake drums 2 and mesh surfaces 3 is installed. To prevent weed impurities for re-cleaning there are protective screens 4 and to remove weed impurities there are weed screws 5. For removal of cleaned cotton and transport to the next technological process in the cleaner there is an outlet pipe 6. In the work of the modernised fine sorghum cleaner, the cotton from the feeder 1 is fed to the stake-plate drums 2 with mesh surfaces 3, where the cotton is cleaned from fine sorghum. Installation of adjacent drums with displacement in the horizontal plane and the opposite direction of their rotation will allow to increase more than 180° angle of coverage of the mesh surface 3 of the stake-slat drum 2, which will lead to a sharp increase in the cleaning effect (in existing cleaners - up to 40%), and zigzag sequential movement of the flow of cotton will provide high reliability of the cleaner in operation.

Also, at each transition from one stake and slat drum to the next, the cotton flow reverses the cleaning surface. This significantly increases the efficiency of cotton cleaning compared to the traditional horizontal arrangement of the fine cleaner's working bodies, where cotton is cleaned unilaterally. When hitting the mesh surface of the next section, the cotton receives an additional impact, which

effectively releases fine weed impurities. A protective screen 4 is installed to prevent weed impurities from entering the secondary cleaning. The weed impurities are removed from the machine by weed augers 5. The cleaned cotton is transported for further processing through the outlet pipe 6.
Economic efficiency of the proposal is formed by increasing the cleaning effect of the equipment, reducing its number in the technological process and, as a consequence, energy consumption, as well as modernised vertical technology of cotton cleaning allows to eliminate the appearance of short fibres in the process of cleaning cotton from small weeds. The analysis of comparative technical characteristics of 1XK brand cleaners and cotton cleaner (Table 3.7) with vertical sections of fine weed cleaning showed that the cleaning effect of the machine increased by 10%, energy consumption decreased by 28.1%.

Comparative technical characteristics of the 1XK cleaner and the cotton cleaner with vertical fines cleaning sections Table 3.7

№	Technological characteristics	1HK	WOK
1	Cleansing effect at initial humidity 8 - 9 per cent and at least 9 per cent weediness	45 - 50	50 - 55
2	Capacity, kg/h, not more: 1st and 2nd grade 3rd and 5th grades	7000 5000	7000 5000
3	Installed power, kW: drive of snare drums power regulator actuator	12 0,25	8 0,25
4	Power consumption at idle speed two with snare drums, kW, not more	1,12	1,12
5	Electricity consumption under load, kWh,	6,4	4,6
6	Air consumption for transporting debris and aspiration, m^3 / s	0,6	0,6
7	Air velocity in the aspiration pipeline, m/s,	18	18
8	Rotational speed, rpm: snare drums feed rollers	420 0 -12	390-42 0 0 -12
9	Technological gaps between the pegs of a cone of drum and mesh, mm	14 - 20	14 - 20
10	Overall dimensions DxWxH, mm	3925x 2670 x1833	1200x 2670 x2000
11	Weight, kg	3100	2800
12	Maximum arc of girth of the four stakes	2744	5832

	of drums with mesh surface, mm		

An industrial sample of a modernised fine litter cleaner with vertically and parallel arranged stake-plank drums with sequential cotton movement is shown in Fig. 3.14.

Metal intensity of the modernised fine sorghum cleaner with vertically and parallel located stake-plank drums with sequential movement of cotton in comparison with the operated fine sorghum cleaner 1HK was reduced by 9.7%, and the maximum arc of circumference of stake drums with mesh surface was increased by 210%, which allowed to increase the cleaning effect of the machine and preserve the natural quality indicators of raw materials.

Fig.3.14 Production sample of modernised fine litter cleaner with vertically and parallel staking and slat drums with sequential cotton movement

VI CHAPTER

TEST RESULTS AND ECONOMIC EFFICIENCY OF IMPLEMENTATION

4.1 Results of production tests of the vertical fine sod cleaner on the UCC unit

The International Cotton Advisory Council (ICAC) predicts that cotton production in 2025 will face challenges related to climate change, high production costs, production technology and policy issues. The two main climate changes that will affect cotton production are rising temperatures and high levels of carbon dioxide concentrations. Rising production costs will require more financial resources, making cotton production a riskier venture for farmers. Cotton agronomic requirements will change dramatically by 2025. Environmental standards will become stricter under pressure from the political sector, civil society and consumers. Cotton farmers will face the need to make very important strategic decisions. Farmers will need to ensure sustainable competitive approaches to balance productivity and fibre quality. Cotton production in 2025 will have to adapt and ensure the coexistence of different forms of production systems such as organic cotton, biotech cotton and fair trade cotton in addition to conventional cotton. Consequently, decision support will need to be reconsidered.

The current direction in the development of new technologies requires mutual coherence between different disciplines. A higher level of multidisciplinary approaches will require better organisation and interpersonal practical experience to guarantee success. Public sector institutions must adapt to the changing atmosphere and reassess their priorities and incentives. However, success depends on investments in technology and viable regulatory systems around the world.

Industrial sample of modernised fine litter cleaner with vertically and parallel arranged stake-plank drums with sequential movement of cotton was manufactured at the production workshops of JSC "Samarkandpaxtamash".

The vertical fine sap cleaner as part of the UCC flow line passed production tests in Kattakurgon Pahta Tozalash JSC at the first unit of the UCC flow line. During the tests comparative data of technical indicators of machines and qualitative indicators of processed cotton were determined.

Comparative indicators of cotton cleaning technology in the in-line UCC with vertical sections of cleaning from fine sap (Table 4.1) showed that the number of stake and plate drums in the modernised unit decreased by two units, the area of "live section" of the mesh surface for drums increased by 38.2%, energy intensity decreased by 11% and metal intensity by 5%.

Due to the modernisation of the fine sap cleaner with vertically and parallel located stake and plate drums with sequential cotton movement in the cotton cleaning unit of the UCC the number of zones of counter rotations of adjacent stake and plate drums was reduced by 41.7%, which allowed to reduce the departure of fibrous material in the weed waste by 32.6% (relatively).

Comparative indicators of cotton cleaning of selection varieties "Bukhara 102", III - industrial grade, 1-class and "Omad", III - industrial grade, 1-class are shown in Table 4.2.

Comparative technological indicators of cotton cleaning technology in an inline line with vertical sections of cleaning from fine sap

Table 4.1.

№	Indicators	Unit variants	
		Operated 2 UHC(4)+1HC units	With two vertical cleaners
1	Number of collar drums in the machines, pcs.	24	22
2	Machine capacity, tonnes/hour	7	7
3	Area of "live section" of mesh surface for cone drums, m^2	24 x 2 x 0.628 = =25.1	14 x 2 x 0.628 = 17.6 8 x 2 x 1.44 = 23.0 17,6 + 23,0= 40,6
4	Energy intensity of machines, kW	128	114
5	Metal intensity of machines, tonnes	40	38
6	Occupied volume of the machine, m^3	20x2x2x2= 80	12x2x2x2 =48
7	Area of drying and cleaning shop, m^2	42x18=756	18x18=324
8	Cost of equipment, thousand UZS	49900x9449x2= 943010	41150x9449x2=777653
9	Number of counter-rotation zones of adjacent peg-bar drums	12	7
10	Short fibre separation during cleaning, kg/hour	Average 1.81 or more	Not more than 1.22

Comparative performance of production tests of a modernised fine litter cleaner with vertically and parallel staking and slat drums with

sequential cotton movement

Table 4.2.

№	Indicators	Options	Indicators of cotton of selective and industrial varieties	
			Bukh-102, W-grade, 1st class.	Omad, W-grade, 1st class.
1.	Initial cotton moisture content, %	Existing	13,5	13,2
		Proposed	13,5	13,2
2.	Initial cotton contamination, %	Existing	3,1	4,6
		Proposed	3,1	4,6
3.	Of which, content of fine sap in cotton, %	Existing	2,4	3,6
		Proposed	2,4	3,6
4.	Cotton moisture content after cleaning, %	Existing	8,7	7,9
		Proposed	8,9	8,1
5.	Gap after cotton cleaning, %	Existing	1,45	2,16
		Proposed	1,12	1,43
6.	Contamination and sum of defects in fibre, %	Existing	1,51	2,52
		Proposed	1,23	2,12

Analysis of the table shows that in the proposed variant of cleaning, the cotton is cleaned up to 33.8% more efficiently than in the existing one. Fibre performance improved up to 18.5%.

4.2 Calculation of economic efficiency from the introduction of a vertical cleaner on the CCS unit

Economic efficiency was calculated mainly taking into account the elimination of short fibres in the process of cotton cleaning from fine weeds and reduction of energy costs. Economic efficiency from the introduction of the modernised vertical cotton cleaner from fine weeds was calculated from comparative data obtained during laboratory and production tests of the machine. According to the passport data the cleaning effect of the serial cotton cleaner from fine weed impurities mark 1 XK at moisture content of 8-9% and cotton clogging less than 9% is 45-50%. Electricity consumption is 12 kW for 8 staking drums.

The gradual and preferential transition to machine harvesting of raw cotton requires special attention in the technology of primary cotton processing, mainly cleaning of the raw cotton. As a result of previous research, a universal cotton complex (UCC) was created at the end of the last century. Along with the advantages of the operating cleaning machines of the UCC in a set with a fine sap cleaner 1 XK, the fine sap cleaning sections work inefficiently, as the working angle of contact between the cone drum and the mesh surface is not more than 100° .

Due to the counter-rotation of the adjacent drums in the area where the cotton meets the second drum, there is considerable damage to the cotton (especially in the case of low grades) due to the abrupt change of direction of the cotton in the next cleaning section in the process. As a consequence, significant damage to fibre and seeds occurs. There are also a number of such problems in the operation of the UCC unit as: high power consumption of the unit - 98 kW, high metal consumption of the structure up to 20 tonnes in a complete set with a 1XK cleaner, frequent slaughtering situations in the processing of low grades, problems in maintenance and cleaning of the unit, a source of emission of fibrous waste into the atmosphere.

In the proposed modernised cotton cleaner from fine litter with vertical compression of stake and plate drums unidirectional speed of rotation of the stake drums allows to eliminate slaughtering situations in the machine. At the same time, the angle of girth of the stake drum with the mesh surface is more than 180° and the shockless trajectory of cotton movement during its cleaning is realised. The given scheme of arrangement of cleaning sections allows to increase considerably the cleaning effect, and also preserves natural qualitative indicators of cotton and its components, that is the basis for development of technology of vertical cotton cleaning at cotton mills.

Calculation of economic efficiency from the introduction of new equipment and technology at cotton ginning enterprises is carried out on the basis of the developed methodological guidelines [45 - p.34].

For comparison of basic and proposed technologies of cotton cleaning from fine weed impurities and economic analysis of the obtained results in the calculations are taken technological indicators of the 1HK cleaning machine operated in the industry and the proposed technology of vertical cleaning of cotton from fine weeds.

Calculation of annual economic efficiency is made on the basis of comparison of expenses for the existing and proposed technology of cotton cleaning from small weed impurities for low grades of cotton, which for an average cotton plant with processing of 30000 tonnes is 26% or 7800 tonnes of the total volume of cotton harvest. Assuming an average fibre yield of 32%, the volume of low grade fibre produced is 2496 tonnes.

On the basis of the above methodological guidelines, we calculate the annual economic efficiency from the introduction of new technology, means of mechanisation and automation of production processes according to the following formula [46 - p.468]:

$$Э = [(C1+E_{н}*K1) - (C2 + E_{н}*K2)]*A + (C_{т}2 - C_{т}1) \quad (4.1)$$

Here:

E - annual economic effect, thousand UZS.

C1 and C2 - current value of equipment of existing and modernised equipment, thousand UZS;

En - normative coefficient of capital investments efficiency - 0.15; K1 and K2 - relative capital investments in the existing and proposed variant of cotton cleaning technology from weed impurities ths.sum;

A- annual volume of output in physical terms;

St1 and St2 - cost of output in the existing and proposed variant of technology of cotton cleaning from weed impurities thousand soums;

Table 4.3 summarises the comparative data for the calculations.

We calculate the parameters to be changed.

Costs for the production of CCS line with vertical cleaners of cotton from fine dust make 82560 thousand soums.

Calculation of capital expenditure

In the basic capital investments in the basic and implemented variants the cost of equipment - 136000 thousand UZS and 82560 thousand UZS is taken into account.

Additional capital expenditures include transportation and installation of equipment (10 % of equipment cost), i.e. 13600,0 thousand UZS and 8256,0 thousand UZS.

Total capital investments including additional capital expenditures are in both variants:

K_1 = 136000 + 13600 = 149600 thousand soums

K_2 = 82560 + 8256 = 90816 thousand soums

Summarised comparative data for the calculations

Table 4.3.

№	Indicators	Unit of measurement	Options	
			Existing technology	Proposed technology
1	Cotton cleaner to remove small weed impurities	pieces	2	2
2	Medium gin machine performance	tonne	7	7
3	Annual working capital of the cotton mill's time	Hour	861	861

4	Annual output cotton fibre	tonne	2496	2498
5	Electricity consumption for in-line CCS complete with cleaner 1XK.	kW/soat	64	57
6	Electricity price per 1 kWh	Sum	348	348
7	Installation price	thousand soums	136000	82560
8	Weight of the flow line CCS complete with cleaner 1XK	tonne	20	12

Calculation of operating costs

Calculation is carried out by variable items and consists of calculations of amounts of deductions for current repairs (Table 4.4).

The cost of current repairs is assumed to be 5% of the equipment cost.

Current repair costs in the base case are as follows:

for current repair: 13600 - 5 / 100 = 6800 thousand soums;

Costs for current repairs in the implemented variant are as follows:

8256 - 5 / 100 = 4080 thousand soums;

Comparative calculation of electricity costs

Electricity costs are calculated according to the following formula:

$$W = Py*Kc*To*Cэ \qquad (4.2)$$

Here, Py - installed power of electric motors;

Kc - demand coefficient;

To - operating time of the equipment;

Sa - price of 1 kWh of electricity;

The cost of electricity in the existing variant is:

W=64*0,7*861*348 = 13423 thousand soums;

As proposed:

W=57*0,7*861*348 = 11955 thousand soums.

Comparative costs of the existing and proposed equipment options

Table 4.4.

№	Name of expenditure	Cost of expenses, thousand UZS	
		Existing variant	Proposed variant
1	Amortisation expense	22 440	13 464

2	Expenditure on current repairs	6 800	4 080
3	Electricity costs	13423	11955
4	Bottom line:	42663	29499

Calculation of economic efficiency due to reduced output fibre to waste

The calculation of economic efficiency is made by the difference of the amount of fibre produced by the existing and proposed technologies of cotton cleaning from weed impurities.

The modernised vertical cotton cleaning technology eliminates the appearance of short fibres during the cotton cleaning process from small weeds.

Loss of fibre for low grades in the process of cleaning from fine sap in the existing technology was 1558 per year (861 soat*1.81 kg of fibre/hour). To carry out calculations, prices for medium-fibre cotton of III grade, III class "urt" with the price of 8084860 soums per tonne of production (Price list of prices for cotton fibre No. 40-02-04-2019, approved by the Ministry of Finance of the Republic of Uzbekistan on 23 October 2019) were adopted [47].

In making the calculations, the average fibre yield of an average cotton ginning mill with a cotton stock of 30000 tonnes was assumed to be 9900 tonnes with an average fibre yield of 33.5%.

We calculate the cost of cotton fibre in the existing and proposed technologies of cotton cleaning from weed impurities:

St1 = 2496* (8084860*1,15) = 23206782 thousand soums.

St2 =2498* (8084860*1,15) = 23225377 thousand soums.

Based on the results obtained, we calculate the annual economic efficiency from the introduction of modernised vertical cotton cleaning technology:

E1 = [(C1 + E n* K 1)- (C 2 + E n* K 2)]*A + (St 2 - St 1) =

= [(136000+ 0,15*149 600) - (82560+ 0,15*90816)] *1 +

+ (23225377- 23206782) = 80853 thousand soums.

GENERAL CONCLUSIONS AND RECOMMENDATIONS

Analysis of research on improving the efficiency of fine sorghum cleaners has shown that a significant disadvantage of existing machines of horizontal layout is that the cotton is subjected to multiple deformation due to countercurrent counter-impact of the staking drums, which contributes to the appearance of short fibres in the composition of weed impurities and significant loss of fibre.

Proceeding from these positions, it is necessary to carry out researches on creation of effective resource- and energy-saving vertical technology of cotton cleaning from fine weeds by increasing the coefficient of efficiency of "live" cross-section of the mesh surface, which would allow maximum preservation of natural qualitative indicators of cotton. It is revealed that at horizontal technology of cotton cleaning from small weed impurities the efficiency coefficient of "live" section does not exceed the value of p =0,25^0,30, while at vertical technology of cleaning this indicator is p = 0,58: 0,60, which is the basis of effective cotton cleaning from small weed impurities.

In a set of vertical cotton cleaner from small weed impurities in the number of four stake drums acting forces on the cotton fly and the rotation frequency of the stake drums increase in the following sequence $F1<F2<F3<F4$ and $v1<v<v_{23}<v_4$, which is the reason for highly efficient vertical technology of cleaning cotton from small weeds. At cleaning of cotton from fine weed impurities on existing horizontal technology of cleaning, speed of rotation of stake drums makes, 9 m/sec. At vertical technology of cotton cleaning there is a possibility of increasing the speed of rotation of staking drums up to the value of 9.67 m/sec. This is due to the fact that there is no counter impact of the stakes drums, present in the horizontal technology of cotton cleaning, which negatively affect the natural quality indicators of processed cotton.

In connection with the absence of counter shock effects on the processed cotton, the loads on electric motors were considerably reduced, as a result of which instead of the total power at horizontal technology of cotton cleaning from small weeds $W = 11$ kW*hour, at vertical technology of cleaning was $W = 6$ kW*hour. The increase in the number of stakes involved in the cleaning process and the increase in the useful area of the mesh surface at the horizontal method of cotton cleaning on the basis of theoretical analysis was 36.2%, at the vertical method of cleaning 37.2%. At values of *friction* coefficient $f=0.1$ and $f=0.2$ respectively at the horizontal cleaning method was 33.06% and 52.0%, at the vertical cleaning method was 43.86% and 69.87%.

It has been determined that the arrangement of staking drums on two vertical parallel axes allows to change the cleaning side to the opposite one when moving cotton from one drum to another. This is one of the components of increasing the cleaning effect of the machine. The obtained equations allow to determine the cleaning effect of the cleaning sections between the stakes of the staking drum. The maximum cleaning effect is achieved in the sections between the first and the third stakes of the drum, in other sections there is a decrease in the cleaning effect. According to theoretical calculations, the increase in the number of *M* leads to a significant increase in the cleaning effect in the first and second cleaning section and can reach 2025%.

It was determined that for the coefficient of *friction of* cotton on the mesh surface equal to $f=0.1$ for the existing technology of cleaning from fine sap (horizontal arrangement of stakes drums) cleaning effect was 33.1%, and at the coefficient of *friction of* cotton on the mesh surface equal to $f=0.2$ cleaning effect was 51.9%. By increasing the number of stakes and the coefficient of friction between the cotton and the mesh surface, the theoretical cleaning effect increased for horizontal cleaning technology to 36.2% ha and for vertical cleaning technology to 37.2%. Comparative indicators of cotton cleaning technology in the in-line UCC with vertical sections of cleaning from fine sap have shown that the number of stake-and-plank drums in the modernised unit decreased by two units, the area of "live section" of the mesh surface increased by 38.2%, due to modernisation in the UCC unit the number of zones of counter rotations of adjacent stake-and-plank drums decreased by 41.7%, which allowed to reduce the departure of fibrous material in weed waste by 32.6% (relatively). In the offered variant of cleaning cotton is cleaned from small weed impurities up to 33,8% more effectively than in the existing one. Due to this contamination and the sum of defects in the fibre decreased to 18.5%. Economic efficiency from introduction of the modernised cleaner with vertical arrangement of sections of cleaning from fine sorghum at the drying and cleaning shop of UCC made 764460 thousand soums per year.

LIST OF REFERENCES

1. Declaration of the participants of the 8th Meeting of the Asian Cotton Research and Development Network. Tashkent. 11.09.2019.
2. Presidential Decree No. UP-60 "On the Strategy for the Development of New Uzbekistan for 2022-2026" dated 28 January 2022.
3. V. G. ARUDE, Cotton ginning ('rechnology, Trouble shooting and maintenance), Indian council of agricultural research, Wadi Nagpur, 2008, p.42-51.
4. http://www.sdmj.com.cn/ey/products.
5. R.A.Gulyaev, A.E.Lugachev, H.S.Usmanov Modern state of production, processing and quality of cotton products in the leading cotton-growing countries of the world: Monograph. Printing house of JSC "Paxtasanoat ilmiy markazi", Tashkent, 2017, - p.11.
6. https://findpatent.ru/patent/200/2004635 .html
7. Lugachev A.E. Research of the basic elements of raw cotton cleaners with the purpose of increase of qualitative indicators of process: Cand.Sci: - Kostroma, 1981. - c.31-32.
8. S.D. Boltabaev Preliminary cleaning of raw cotton of machine harvested cotton from weed impurities: Cand. Candidate of Technical Sciences: - Tashkent, 1949, - p.156.
9. Budin E.F. Research of grates-sawing working bodies of cleaners of raw cotton of machine harvested cotton of medium-fibre varieties: Cand. Candidate of Technical Sciences: - Tashkent, 1968, - p.146.
10. G.D.Jabbarov Primary processing of cotton: "Legkaya Hindustriya", Moscow 1978, - p.124.
11. Musakhodjaev Z.M. To the question of creation of combined cleaner of raw cotton of machine harvesting, : Dissertation ... Candidate of Technical Sciences: - Tashkent, 1970, - p.11.
12. V.A.Bogomolov Research and selection of technological process of raw cotton cleaning of machine harvested cotton in the Republic of Azerbaijan, : Cand. Candidate of Technical Sciences: - Kirovabad, 1974, - p.171.
13. A.L.Sapon Research and development of technological progress of primary processing of raw cotton of machine harvest on the basis of full-process flow line: Cand. Candidate of Technical Sciences: - Tashkent, 1978, - p.135.
14. Abduazimov Sh.H. Increase of efficiency of raw cotton purification from small weed impurities by improving grates: Cand. Candidate of Technical Sciences: - Tashkent, 1997, - p.134.
15. Lugachev A.E. Development of theoretical bases of feeding and cleaning of cotton in relation to the in-line technology of its processing: Cand. Doctor of

Technical Sciences: - Tashkent, 1998, - p.99 - 136.
16. Bobomatov A.H. Creation of an effective design and improvement of scientific bases of methods of calculation of the cotton cleaner from fine sap: Cand. Cand.tehn.nauk: - Tashkent, 2017, - p.115.
17. I.D.Madumarov Increase of efficiency of cleaning process on the basis of optimisation of heat-moisture condition and uniform feeding of cotton: Diss... Doctor of Technical Sciences, 2019, - p.181.
18. D.A.Usmanov, R.H.Karimov, K.K.Polotov Technological evaluation of four-drum cleaner operation. Problems of modern science and education, Ivanovo, 2019, http://http://www.ipi1.ru.ipi1.ru.
19. Usmanov H.S., Ismailov A.A., Makhmudov Y.A. Calculation of force effects on cotton fly at horizontal cleaning technology// Materials of the XVI International scientific and practical conference Cutting-edge science - 2020 , April 30 - May 7, 2020: Baku. EDUCATION AND SCIENCE Ltd -233 p.
20. A.M. Aboukarima, H.A. Elsoury and M. Menyawi. Artificial Neural Network Model for the Prediction of the Cotton Crop Leaf Area. Agricultural Engineering Research Institute, Agricultural Research Centre, Dokki, Giza, Egypt.
21. https://www.colorcodepicker.com/
22. Code of Federal Regulations (CFR). 2010. Method 201A-Determination of PM10and PM2.5emissions from stationary sources (Constant sampling rate procedure). 40 CFR 51, Appendix M. Available at http://www.epa.gov/ttn/emc/ promgate /m-201a.pdf (verified 19 Aug. 2013).
23. Environmental Protection Agency (EPA). 2010. Frequently asked questions (FAQS) for Method 201A [Online]. Available at http://www.epa.gov/ttn/emc/ methods/method201a. html (verified 19 Aug. 2013).
24. National Agricultural Statistics Service (NASS).1993-2012. Cotton Ginnings Annual Summary [Online]. USDA National Agricultural Statistics Service, Washington, DC Available at http://usda.mannlib.cornell.edu/MannUsda/ viewDocumentInfo.do?document ID=1042 (verified 19 Aug. 2013).
25. Valco, T. D., H. Ashley, J. K. Green, D. S. Findley, T. L. PriceJ.M. Fannin, and R.A. Isom. 2012. The cost of ginningcotton-2010 survey results. p. 616-619 In Proc. Beltwide Cotton Conf., Orlando, FL. 3-6 Jan. 2012. Natl. Cotton Counc. Am., Memphis, TN.
26. Whitelock, D.P., C.B. Armijo, M.D. Buser, and S.E. Hughs.2009 Using cyclones effectively at cotton gins. Hughs.2009 Using cyclones effectively at cotton gins. Appl. Eng. Ag. 25:563-576.
27. Armijo, C.B., and M.N. Gillum. 2010. Conventional and highspeed roller

ginning of upland cotton in commercial gins.Appl. Eng. Agric. 26:5-10.
28. Kh.S.Usmanov, A.M.Salimov, F.N.Sirozhiddinov Increase of the Efficiency of raw cotton cleaning and analysis of factors influencing this process. Proceedings of XXIV International Scientific and Practical Conference "EurasiaScience", 14.10.2019, - p.64-65.
29. H.S. Usmanov, I.Z. Abbozov, A.T. Doliev Tozalash zharayonida kozikchali barabalarni pakhtaning tabiyi khusususiyatlariga tajsirining nazariy tahlili // "Mexanika muammolari" journal No. 3, 2019 yil, 60-63 betlar.
30. Usmanov Kh.S., Sabirov I.K., Khaitbaev Kh.Kh. Calculation of impact effects on cotton fly under existing horizontal cleaning technology // International scientific and practical Conference Modern views and research - 2021, January-February, 2021: Egham.Independent Publishing Network Ltd -14, - pp. 69-74, DOI: . 69-74, DOI: http://doi.org/10.37057/E_7/
31. Usmanov Kh.S., Sabirov I.K., Khaitbaev Kh.Kh. Calculation of force effects on cotton fly during cleaning on vertical cleaner // International scientific and practical Conference Modern views and research - 2021, January-February, 2021: Egham.Independent Publishing Network Ltd -14, - pp. 74-78 DOI: http://doi. org/10.37057/E_7.
32. Usmanov Kh.S.,Salimov A.M.,Abbozov I.Z.,Doliyev A.T.,Tangirov A.A Theoretical Analysis of the Effect of Spike Drums on the Natural Qualitative Indicators of Cotton at its Cleaning //International Journal of Advanced Research in Science,Engineering and Technology, India, Vol. 6, Issue 6 , September 2019, - pp.67474 - 107474. 6, Issue 9 , September 2019, - pp.10742 - 10747. www.ijarset.com.
33. Usmanov H.S., Alimov M.A., Doliev A.T. Drum kozikchalari pakhta tolasiga tasir etuvchi kuchini aniklash va nazariy takhlili "Mashinashunoslikning dolzarb muammolari va ularning echimi" Academician Kh.Kh. Usmonkhujaev tavalludining 100 yilligiga baFishlangan Respublika ilmiy-amaliy conference makolalar tuplami, Toshkent, 2019 yil 20-21 November, 34-37 betlar.
34. Patil.P.G., Anap G.R., Arude V.G. Design and development of cylinder type cotton pre-cleaner. Agricultural Mechanisation in Asia, Africa and Latin Amer^a. 2014, ISSN: 00845841.
35. Carlos B. Armijo, Kevin D. Baker, Sidney E. Hughs, Edward M. Barnes, and Marvis N. Gillum Harvesting and Seed Cotton Cleaning of a Cotton Cultivar with a Fragile Seed Coat The Journal of Cotton Science 2009. No.13:- pp.158-165).
36. Usmanov H.S., Gulyaev R.A., Lugachev A.E. Search and development of innovative solutions in the issues of effective cleaning of raw cotton Scientific and Technical Journal "Tukimachilik muammolari" 2018 No.3, -C.31.

37. B.M.Mardonov, H.S.Usmanov, F.N.Sirozhiddinov Modelling of raw cotton cleaning process under the action of vertically arranged stake drums //Problems of Mechanics. - 2018, №1.- C 81-86.
38. Usmanov H.S., Abbozov I.Z., Sirozhiddinov F.N. Innovatsion vertikal tozalagichning tozalash samaradorligiga tajsir etuvci omillar tahlili //: "FarFOna polytechnika instituta ilmiy tekhnika journali" 2019 yil, 23-volume 3- soni, 44-50 bet.
39. GOST O'z DSt 643:2006
40. GOST O'z DSt 644:2006
41. GOST O'z DSt 592:2008
42. Adler Y.P., Markova E.V., Granovsky Yu. Granovsky Y.V. Experiment planning in search of optimal conditions. - Moscow: Nauka, 1976. -275 c.
43. Jamalova M.M. To the question of cleaning raw cotton from fine sap. //Dissertation of Candidate of Technical Sciences. Tashkent, 1961, p.77
44. B.M.Mardonov, H.S.Usmanov, F.N.Sirozhiddinov Mechanika muammolari No.1 - 2019, pp.27-32.
45. Methodology for determining the economic efficiency of the introduction of new technology, invention and rationalisation proposals.-M., 1988. p.. 34.
46. Isaev P.A. va boshkalar. Ishlab chikarishni tashkil etish va business reja. "Tafakkur" nashriyoti, Toshkent, 2011, 468 bet.
47. Pakhta tolasining ulgurji narkhlari narkhnomasi. no. 40-02-04-2019. Uzbekistan Respubliki Moliya Vazirligi. 23.10.2019 й.

Printed by Books on Demand GmbH, Norderstedt / Germany